FRAUD AND MISCONDUCT IN RESEARCH

In *Fraud and Misconduct in Research*, Nachman Ben-Yehuda and Amalya Oliver-Lumerman introduce the main characteristics of research misconduct, portray how they are distributed and identify the elements of the organizational context and the practice of scientific research which enable or deter misconduct. Of nearly 750 known cases between 1880 and 2010 which they examine, the overwhelming majority took place in funded research projects and involved falsification and fabrication, followed by misrepresentation and plagiarism. The incidents were often reported by the perpetrator's colleagues or collaborators. If the accusations were confirmed, the organization usually punished the offender with temporary exclusion from academic activities and institutions launched organizational reforms, including new rules, the establishment of offices to deal with misconduct, and the creation of re-training and education programs for academic staff. Ben-Yehuda and Oliver-Lumerman suggest ways in which efforts to expose and prevent misconduct can further change the work of scientists, universities, and scientific research.

Nachman Ben-Yehuda is Professor in the Department of Sociology and Anthropology at the Hebrew University in Jerusalem.

Amalya Oliver-Lumerman is Professor in the Department of Sociology and Anthropology at the Hebrew University in Jerusalem.

Fraud and Misconduct in Research

DETECTION, INVESTIGATION, AND ORGANIZATIONAL RESPONSE

Nachman Ben-Yehuda
and
Amalya Oliver-Lumerman

UNIVERSITY OF MICHIGAN PRESS
Ann Arbor

Published in the United States of America by the
University of Michigan Press
Printed and bound by CPI Group (UK) Ltd, Croydon, CR0 4YY

2020 2019 2018 2017 4 3 2 1

A CIP catalog record for this book is available from the British Library.

Library of Congress Cataloging-in-Publication Data

Names: Ben-Yehuda, Nachman, author. | Oliver, Amalya Lumerman, author.
Title: Fraud and misconduct in research : detection, investigation, and organizational response / Nachman Ben-Yehuda and Amalya Oliver-Lumerman.
Description: Ann Arbor : University of Michigan Press, [2017] | Includes bibliographical references and index.
Identifiers: LCCN 2017020375 | ISBN 9780472130559 (hardcover : alk. paper) | ISBN 9780472123230 (e-book)
Subjects: LCSH: Fraud in science. | Research—Moral and ethical aspects. | Science—Moral and ethical aspects.
Classification: LCC Q175.37 .B46 2017 | DDC 658.4/73—dc23
LC record available at https://lccn.loc.gov/2017020375

"I am that part of the power that eternally wants bad and eternally does only good."

Goethe's Faust

"I am just looking for an honest man."

Attributed to Diogenes of Sinope's (412 [or 404] BC–323 BC) answer to why he was walking in full daylight with a lamp (source: Wikipedia)

Contents

Preface and Acknowledgments

The beginning. Assume you—the reader of this book—hold a managing administrative position at a university, and you get some information accusing one of the researchers in your department of inventing findings or observations or cheating in some other way in publications. What would you do? How would you react? Who would you approach? How far would you go?

This happened to us about ten years ago. We were involved (in our administrative roles at that time) in a case of suspected research fraud that exemplified how complex dealing with such issues can be.[1] The case became a big ordeal for the department, the faculty, and the university. The entire affair, which lasted several months, made us curious. It obviously raised interesting questions about the more general phenomena referred to by the United States Office of Research Integrity (ORI) as "scientific misconduct," or, more simply, fraud in research. It was thus this affair that propelled and motivated us to examine in depth issues of misconduct and fraud in research.

The numerous discussions we had about the event we had to deal with, as well as about other cases that came to our attention, led us to become even more curious about other cases: their characteristics, cultural differences, gender issues, and how they were dealt with by universities and other stakeholders. As a research team, we felt that the combination of a sociologist of deviance (Ben-Yehuda) who had already worked on the issue of fraud in science, and an organizational sociologist (Oliver) could offer alternative and refreshing theoretical foci to the phenomena we wanted to study. And so the idea of this book was born. We began with a small grant and then received additional grants. With three highly dedicated research assistants, we collected information on all cases of detected fraud in research we could find and began to explore the evidence.

Main arguments. This book analyzes fraud in research in a comprehensive and comparative manner and is aimed at a wide readership. It consists of a large assembly of material meant to cover all relevant and important aspects of fraud in research. Because the book uses a wide-angle approach to the topic, it is important to clarify that it does not consist of some haphazard collection of facts but reflects real and tangible arguments; some run throughout the book and some are specific.

While the book focuses on a distinctive form of deviance—fraud in research—the canopy under which this book needs to be thought of is the exploration of deviance within an organizational context. Unlike other forms of cheating, fraud, and conning, fraud in research almost always takes place within an organizational structure such as a university, research institute, or industrial firm. Because of this, we are convinced that examining fraud in research needs to be done with an eye to relevant organizations and stakeholders. This argument runs throughout the book. Another argument is the attempt to guesstimate the magnitude of this form of deviance. It is our conclusion that fraud in research is probably more like an iceberg phenomenon than like a bad apple one. A third argument is that we need to be very careful before applying overly harsh controls to suppress fraud in research. The equation here is that on the one side too much control will probably have a chilling and squelching effect on creativity. On the other side, leaving an independent and autonomous space to researchers will encourage creativity but will also probably open the door to fraud. Is it worth the risk? We believe it is. A fourth argument is that we need to distinguish between deviant scientists—those who commit acts of fraud—and deviant sciences—that is, knowledge claims that are not accepted as legitimate. Fifth, we believe that committing acts of fraud in research involves a process that deviant researchers go through that leads from bona fide, honest, nondeviant researchers to fraudsters. That is, a process of becoming deviant is involved. Sixth, we believe that one of the ways universities managed to cope with issues of fraud in research is by avoiding the rhetoric of "fraud" and instead deflecting the problem to a rhetoric of "integrity" or "ethics." Finally, rather than take one or two cases and "run" with them throughout the book, we opted to use as many illustrative cases as necessary and highlight relevant issues in each of the cases. Being exposed to many cases is—we believe—an interesting and persuasive method to show the size and variance of fraud in research.

We invite you, the reader, to join us in this fascinating tour of examining how and why those who are trusted to supposedly seek a truth subvert this search and betray this trust and loyalty to scientific research. We promise to try to present this highly complex issue in as simple a way as possible so that readers who are unfamiliar with this territory will enjoy a cherished learning experience as well. Moreover, we will open the perspective to view how organizations react to these subversions and violations and how one needs to view all this in a more general context; that is, deviance within organizations. We summarize this work with some conclusions and suggestions.

Acknowledgments. We are grateful for the funding by the Shaine Center and from the Hebrew University. These funds provided the means that enabled us to do this research and some of the editing. We are grateful to the Hebrew University's Authority for Research and Development, whose grant helped in making this book.

It is with great pleasure, respect, and profound gratitude that we extend our hearty thanks and appreciation to our dedicated research assistants—Noga Buber-Ben-David, Shirly Dorfsman, and Ariel Yankellevich—who spared no effort to turn every stone they could to help us create the database used for this study. Philip Jenkins and Gershon Ben-Shakhar kept us updated about various cases and issues, for which we are very grateful. Kennie Lyman's critical and attentive editorial comments are gratefully appreciated. Brian Martin, Kathleen Montgomery, Joseph Lampel and our anonymous reviewers provided useful comments that were integrated—with much gratitude—into the text.

The book was written during the past two years, mainly during summer sabbaticals from the Hebrew University. Part of our intensive discussions took place during the first year at the Chelsmore, and we are grateful to Michael Rahav and the Chelsmore team for helping make this possible. During these summers, Prof. Barbara Weinstein from the Department of History hosted Ben Yehuda in the second year, while the Department of Management and Organizations and the Department of Sociology at New York University hosted Oliver in the first and the second years. We express our deepest gratitude and appreciation for their hospitality.

Many of our colleagues and acquaintances provided access to narratives, criticisms, queries, and numerous discussions about fraud in research, and this consequently helped to crystallize our view regarding what directions universities take and what directions they can take. In fact, whenever we told a colleague about this research project we typically heard about one or more accounts of undisclosed fraud and misconduct. In a way, this experience helped us to reach one of our conclusions that known cases of fraud and misconduct probably represent the tip of the iceberg rather than merely a few bad apples.

We are very grateful and appreciative for the continuous support of our spouses Nir Oliver and Etti Ben-Yehuda in our academic work.

ONE | Introduction

Fraud in scientific research has attracted the attention of scholars and laypersons alike. From debates, warnings, and soul searching in academic and professional journals to op-eds, reports, and headlines in the popular media, many interested parties have tried to dive into this area of deviant behavior. Clearly, researchers who fabricate and falsify data are committing acts of deviance within typically esteemed organizations where they practice their skills. Thus, while examining fraud in research via the lens of deviance gives us an interesting and powerful point of view, it is not a sufficient perspective. The complementary organizational perspective, which informs us how organizations and institutions (e.g., universities) react to, contain, and change when they face acts of fraud, is essential if we are to gain a much better understanding of this curious phenomenon. Consequently, the book that you are about to read takes a different view then previous works in this area. It relies on the expertise of the two authors in deviance research and organizational theory respectively to examine fraud in research as a specific instance of deviance within organizations. We thus intend this book to be of interest to research scientists in all disciplines, people who work for research funding agencies, top administrators at universities, and journal and book editors.

We begin by looking at the nature of the phenomenon to be examined. Then we present and interpret the results of our research effort, through which we located about 750 known cases of fraud in research. The following chapters contextualize our findings in the organizational setting and in relevant formulations in deviance research. This new and integrated view helps us to present a better understanding of, and more insights into, the phenomenon of fraud in research in its most relevant contexts. To help get us into our study, let us first be reminded of some interesting cases of fraud in research in a historical perspective.

ILLUSTRATIVE CASES: A HISTORICAL VIEW

1974. Dr. William T. Summerlin, working at the Memorial Sloan-Kettering Cancer Center in New York City, claimed he had succeeded in transplanting tissues from genetically different mice by a new technology. His proof? He showed white mice with transplanted black patches on them. Upon investigation, it was found that these supposedly transplanted patches were actually painted with simple ink that could be erased with alcohol. The discovery of this fraud was rapid but uncomfortable. British biologist and cowinner of the 1960 Nobel Prize in physiology or medicine, Sir Peter Medawar,[1] whose work focused on graft rejection, was well aware of what was going on and found himself, by his own confession (Zuckerman 1977, 96), "[l]acking in moral courage . . . to say at the time that . . . we were the victims of a hoax or confidence trick." A scientific investigation into this case found Summerlin guilty and he was convicted of misrepresentation. Summerlin himself admitted he had inked the back of a mouse, making this clearly a case of intentional fraud. But Summerlin had been determined to prove that skin grafts between genetically different animals was possible, so this was a clear case of a researcher who resorted to fraudulent practices in order to "prove" something he believed in.[2]

Dr. Summerlin was characterized as suffering from emotional stress and disturbances and was offered a long period of medical leave on full salary ($40,000!) to enable him to rest and obtain professional care.[3]

1978–1980. Wade (1981) reports on John Long, a respected pathologist from Massachusetts General Hospital, who had a United States federal researcher grant of $750,000 and researched Hodgkin's disease. The research involved establishing several human cell lines with traits characteristic of the disease. These lines are difficult to create and maintain, but they are of tremendous value for studying the disease. Stephen Quay, another researcher and colleague of Long, suspected in October 1979 that Long's results were faked. Quay formally expressed his suspicion in a report he prepared in January 1980 accusing Long of reporting on experiments he never actually performed. Under pressure, Long admitted to making mistakes and attributed them (p. 1023) to "pressure for a grant application."[4] He subsequently admitted to the charge that he had forged data. In fact, he fabricated and falsified data on cancer cells between the years 1978 and 1980. He resigned from Massachusetts General Hospital on January 31, 1980, contrite and agreeing not to engage in research any longer. His published work was retracted in the *Journal of the National*

Cancer Institute. Long found later a comfortable position as a pathologist in a midwestern hospital.[5]

1981. John Darsee moved from Emory University, where he was chief medical resident at Grady Memorial Hospital, to Harvard University in 1979. He worked at the Cardiac Research Laboratory there and produced quite a few impressive papers. In May 1981, several researchers at the Harvard Medical School were amazed to witness how Darsee "flagrantly concocted data for an experiment" (Broad, 1982, 478). Consequently, some suspicions about his work were raised, but a Harvard investigating committee that looked into Darsee's works found no reason to be concerned about fraudulent work, although they did find overreporting of work hours. While Darsee's fellowship was terminated, the National Institutes of Health (NIH) (who funded the research) was not informed about this. As more suspicions and research discrepancies were brought up, the National Institutes of Health began a formal inquiry. The results of this inquiry were that Darsee had fabricated data and reported on experiments that had never been carried out. Along the way, the investigation conducted by Harvard University was criticized as shallow and inadequate. As the investigation against Darsee continued, it found that when Darsee was an undergraduate he had already falsified research data. Researchers who published with Darsee were criticized for signing their names on papers they were unfamiliar with.[6]

Claudia Wallis (1983)—a journalist for *Time*—pointed out that Darsee claimed to have had no recollection of committing a scientific fraud. In the "Correspondence" section of the prestigious *New England Journal of Medicine* Darsee wrote that:

> I am deeply sorry for allowing these inaccuracies and falsehoods to be published in the Journal and apologize to the editorial board and readers of the New England Journal, to Emory University, and to my coauthors. Dr. Nutter and Dr. Heymsfield are impeccably honest researchers whose names appeared on these papers in good faith. Neither of them was aware of any of the inaccuracies nor were they responsible for any of them.[7]

1981. In 1981 a committee of the U.S. Congress began to raise questions about the money spent by the National Cancer Institute (NCI) in its "war on cancer." One of the cases the committee wanted to look into was

> the charge that the N.C.I. awarded a $910,000 research grant to a researcher after he was forced to resign from Boston University amid charges that he had falsified data and injected patients with unproven toxic drugs without proper consent under another cancer agency program. The N.C.I. did not begin its own investigation of the alleged fraud of the researcher, Dr. Marc J. Strauss, until after the *Boston Globe* exposed the affair last June, two years after Dr. Strauss's resignation, and Dr. DeVita[8] withheld information about allegations from the peer review committee of researchers who approved the new grant in January of 1980 and from the National Cancer Advisory Board, a 23-member panel that oversees the agency.
>
> In a letter to Mr. Hatch,[9] Dr. DeVita said it would have been "improper" to inform the panels of the charges hanging over Dr. Strauss, now with the New York Medical College and the Westchester County Medical Center, because they had not yet been proven and the new grant, on cell kinetics, involved no clinical work with patients.[10]

Journalist William J. Broad added (1981) that Marc Strauss, a researcher from Boston University, was accused in 1978 by his twenty-person staff of submitting falsified reports. Strauss resigned, insisting that he was the victim of a conspiracy planned and executed by his staff. Working hard again, Strauss re-established himself and applied to the National Cancer Institute (NCI) for a new grant and was given a $1.32 million grant. When the NCI board learned of Strauss's past, there was an uproar that initiated a congressional investigation. During the inquiry it was made clear that the National Institutes of Health (NIH) would neither allow the creation of blacklists nor let allegations prevent anyone from receiving support. In 2000 it was reported that Strauss became a poet and practiced medicine in New York.[11]

1986–1998. The case that never was: the Thereza Imanishi-Kari–David Baltimore affair. This famous case rocked the halls of science, producing publications for almost a decade between around 1986 and 1998. Except for the lack of romantic involvements, it has much of the drama and complexity of a "whodunit" detective movie. Quite a few articles and books were written on this affair, and delving into the details can easily take a whole new book. Let us summarize this affair briefly, based mostly on succinct Wikipedia summaries.[12]

In 1986 David Baltimore, then a faculty member in biology at MIT and previous (1975) cowinner of the Nobel Prize in physiology or medicine (with two colleagues), shared with four others a scientific publication with Thereza Imanishi-Kari (then a professor of biology at MIT) in the prestigious journal *Cell*. Margot O'Toole, who had a postdoctoral position with Imanishi-Kari (and was not one of the authors of the paper) raised concerns about the paper and eventually accused Thereza Imanishi-Kari of fabricating data in a cover-up. This opened a Pandora's box. As O'Toole assumed the role of a whistle-blower, the ball of suspicion continued to roll. Two functionaries at NIH—Walter W. Stewart and Ned Feder—began an investigation, and Rep. John Dingell (D-MI) also exerted pressures, which culminated in the involvement of U.S. Secret Service document examiners. In 1990 David Baltimore became the president of Rockefeller University. He had to visit quite a few laboratories there and listen and respond to concerns aired by faculty-activists about the affair. Nevertheless, in the midst of the affair, he resigned from this prestigious position. In a draft report dated March 14, 1991, and based mainly on the U.S. Secret Service forensics findings, the Office of Scientific Integrity (OSI) accused Imanishi-Kari of falsifying and fabricating data and criticized Baltimore for failing to respond properly to O'Toole's suspicions. It did not take long for the report to leak to the media. Two events followed this. One was that David Baltimore and three coauthors retracted the paper. Thereza Imanishi-Kari and Moema H. Reis, however, did not sign the retraction. Then, in 1992 the U.S. attorney for the District of Maryland, who had been investigating the case, announced he would bring neither criminal nor civil charges against Imanishi-Kari. Nevertheless, relying mostly on the analysis of laboratory notebooks by the Secret Service, in 1994 the Office of Research Integrity (ORI) found Imanishi-Kari guilty on nineteen counts of research fraud. But that was not the end. In 1995 an appeal panel of the Department of Health and Human Services reviewed the charges and in 1996 concluded that the ORI had failed to prove even one of its nineteen charges and dismissed all charges against Thereza Imanishi-Kari. Throughout all these events, David Baltimore was never suspected of any misconduct.

The years in which this affair exploded were years when attention to and concerns about scientific fraud were at their peak, so it is not difficult to understand the turmoil around such an affair in a turbulent and volatile atmosphere. This affair illustrates how difficult and nerve-racking an investigation of a suspected fraud case can be.[13] Moreover, this case brought to the forefront the issue of whistle-blowers in such cases.[14]

1998–2015. These seventeen years have witnessed a gloomy case involving Dr. Andrew Jeremy Wakefield. In 1998 Dr. Wakefield published a paper in the British medical journal the *Lancet* in which he claimed that the administration of the measles, mumps, and rubella (MMR) vaccine was connected to the appearance of autism and bowel disease. Replications failed to repeat the results, and an inquiry by the British General Medical Council in January 2010 found a series of professional and ethical problems in the study. Consequently, the *Lancet* retracted the 1998 paper. Because the 1998 paper was false, Dr. Wakefield lost his license to practice medicine in the United Kingdom. Unfortunately, too many parents believed Wakefield's false claims, and vaccination rates in some countries declined, leading to an increase in the occurrence of measles and mumps and many fatalities. Sadly, this bogus study also contributed to an atmosphere of mistrust in vaccines more generally.[15] Examining the Disneyland measles outbreak, the January 23, 2015, edition of the *Washington Post* echoed an editorial text published earlier in the *Los Angeles Times* that blasted the antivaccination movement.[16] What makes this case particularly interesting is that a fraudulent work was able to convince enough people not to take vaccines that the vaccination rate was lowered, which resulted in widespread misery and tragedy.

2002. Perhaps the most famous case was the one involving German physicist Jan Hendrik Schön.[17] He became famous following what he reported as significant breakthroughs in semiconductor research. As suspicions grew about the work, a 2002 committee appointed to investigate the case found that his work was based on fraud. The case created many debates on the role and responsibilities of peer reviews.[18]

2006–2009. Hwang Woo-Suk was a South Korean veterinarian researcher and a professor at Seoul National University (dismissed on March 20, 2006) who became infamous for fabricating a series of experiments in the field of stem cell research.[19] On May 12, 2006, Hwang was "indicted on embezzlement and bioethics law violations linked to faked stem cell research." *Korea Times* reported on June 10, 2007, that the university had expelled Hwang; the government rescinded its financial and legal support and barred him from conducting human cloning research. In a sign of things to come, Hwang was sentenced at the Seoul Central District Court on October 26, 2009, to a two-year suspended prison sentence after being found guilty of embezzlement and bioethical violations but cleared of fraud.[20] In later years Hwang tried to begin a new career in cloning.[21]

2005–2010. Perhaps the most dramatic case in recent years was the affair involving Eric T. Poehlman from the University of Vermont, who specialized in the field of aging and obesity. Following suspicions of fraud in his work, he admitted in March of 2005 that he had falsified seventeen grant applications to the National Institutes of Health to make his work seem serious and promising and help him win nearly $3 million in government funding. In addition, he fabricated data in ten of his papers that were submitted between 1992 and 2000. In June 2006, Poehlman was convicted and sent to prison for one year and a day, the first known case of a researcher who stood trial and went to prison after fabricating and falsifying research.[22]

A prison sentence was also given in April 2013 to Steven Eaton "a UK researcher who cooked experiments while at the U.S.-based contract research outfit Aptuit." He was "given a three-month prison term, making him the first person to serve time under a 1999 British law called the Good Laboratory Practice Regulations"[23]

Another famous case is that of Tufts University specialist and leader in managing pain Scott S. Reuben. Following suspicions, it turned out that while Reuben's work was based on zero clinical trials, it had influenced the way physicians treated pain for years. In May 2010, he was sentenced to a prison sentence of six months plus heavy fines.[24]

2012. Many of us are under the impression that drinking red wine may have beneficial effects on our heart's health. Following an anonymous tip, the University of Connecticut appointed a review board that examined the research done on exactly this issue by one of its faculty, Dipak K. Das, who at the time was the director of the Cardiovascular Research Center at the University's Health Center. This process began in 2009, and in 2012 the review board pointed out in a report that it suspected that Das's work involved "145 instances of fabrication and falsification of data. Many involved cutting and pasting photographic images."[25]

2013. This year had a few cases. One was a case of a chief of a medical department who was charged with fraud in the context of a scientific service she was providing. Most cases of fraud do not cause misery to a large numbers of victims, but 2013 witnessed a case that gave us a very different perspective. According to a report in the *New York Times* (August 20), Annie Dookhan, a thirty-five-year-old chemist who worked for about nine years in a Massachusetts state crime lab, was accused of falsifying the results of forty thousand drug tests (and possibly more) that were brought to her for inspection. Her job was to use proper chemical testing to determine if the samples that were brought to her contained controlled substances. The report suggested that Dookhan was probably driven by ambition to excel, and the method she used to fake the evidence was simple enough; she just looked at the samples without performing any chemical analysis. Moreover, in some cases she altered the samples to cover up her falsifications. Thousands of people were punished based on falsified evidence. While a public report about this case was made in the summer of 2013, Dookhan was prevented from performing any drug testing already in June 2011 and later resigned. The *New York Times* stated that Dookhan was arrested and "charged with lying about the integrity of drug evidence that she analyzed."[26] After pleading guilty in 2012 to falsifying thousands of drug tests, Dookhan was sentenced in 2013 to at least three years in prison.[27]

2014. Early this year, Japanese cell researcher Haruko Obokata headed a team of researchers who published two papers in *Nature* claiming that stem cells could be created by exposing mature cells to stress. Nothing of the sort had ever been achieved prior to these publications, so obviously these papers drew much attention. A committee was set up to examine these unexpected results, and its first results were not consistent with the claims. One might infer that fraud was suspected, but in a press conference in March 2014 it was announced that while no fraud had occurred, inappropriate handling of data was involved.[28] Nevertheless, the June 14, 2014, issue of *The Economist* reported that the committee had concluded that in this case data were manipulated in a way that classified this research as fraud. Unfortunately, this case involved a tragedy. On August 5, 2015, Reuters reported that fifty-two-year-old Yoshiki Sasai, "co-author of the high-profile research that had seemed to offer hope for replacing damaged cells or even growing new human organs, was found [dead] early on Tuesday at the Riken institute where he worked in Kobe, western Japan, police and the institute said. 'It is confirmed as a suicide,' said a police spokesman. 'It was a hanging.'"[29]

A less dramatic event took place in 2014 when [t]he U.S. Army War College stripped U.S. senator John Walsh, a Montana Democrat, of his master's degree. "Walsh's office said Friday that the college had revoked his degree following an investigation into plagiarism allegations regarding a 2007 paper he wrote while he was student there. The allegations came to light this summer, when the *New York Times* reported that large chunks of his paper had been lifted from other sources, without proper attribution. Walsh said he disagreed with the college's findings, but accepts the decision."[30] As plagiarism is one of the most common forms of deceit in research, the Walsh case should not really surprise us.[31]

2015. This year witnessed some spectacular cases. Suspicions regarding fraudulent studies of the implications of same-sex marriages[32] surfaced and caused significant discussions in the media. Later that year an Iowa State University biomedical researcher—Dong-Pyou Han—who faked HIV vaccine trials was sentenced to four and a half years in jail plus a fine of $7.2 million to be paid to the NIH, the federal government agency that funded the research.[33] The year has also witnessed David E. Anderson—a graduate student in the Department of Psychology at the University of Oregon—admitting to have faked data in four neuroscience papers. Prof. Edward Awh, a brain researcher who was his former supervisor, requested that all papers be retracted.[34] Another tempest not only raised some doubts about an ethnographic book by Alice Goffman but also generated critical notes regarding the entire review process for sociology texts.[35]

Another case involved fraudulent cancer research. Following the completion of his training in hematology and oncology at Duke University in 2006, Anil Potti began to write about a new way to choose the best chemotherapy treatment. He claimed he could pinpoint the best chemotherapy for patients based on their cancer's genetic signature. But on Monday, November 9, 2015, some nine years after his first paper was published, government investigators from the Office of Research Integrity (ORI) announced that his research was mostly made up. He also repeatedly represented himself as a Rhodes Scholar in grant applications although he had never actually won the prestigious scholarship.[36] At least eleven of Potti's papers have been retracted, and he has been reprimanded by the medical boards of both Missouri and North Carolina. As of 2012, Potti was a practicing oncologist in North Dakota.[37]

Cases continued to be exposed during 2016. One famous case was the decision by the University of Tokyo to launch an investigation due to suspicion that data were manipulated by research groups (Normile 2016).

THEN AND NOW: CHANGES IN THE MAKING

A long time passed between the case of Summerlin, who was sent to recuperate and rest on medical leave, Hwang, who was given a suspended sentence, and Poehlman, Eaton, Reuben and Han, who were sent to prison. The abyss separating these cases is not only measured by years but also profoundly by a significant change in moralities, norms, and attitudes and organizational cultures. Conducting scientific research and the organizational accountability and reaction to cases of fraud have changed, as the different cases of fraud indicate.

Cases of fraud in research are numerous. The database of cases involving fraud in research we have created consists of a few hundred such puzzling and intriguing cases. The previous illustrative cases demonstrate the puzzle that we set out to solve: how and why researchers fabricate, cook, massage, and plagiarize data and how the organizations within which they commit these acts cope and process such behavior. Trying to make sense of cases where researchers—supposedly seekers of truth—are involved in deliberate attempts to create a false reality, persuade audiences that these realities are valid, and enjoy the benefits that accompany scientific revelations, whether those of prestige, money, reputation, or power, is not an easy or simple exercise.

Nevertheless, our quest to explore and solve this riddle of deviance within organizations led us into some mesmerizing and challenging alleys, as well as to a realization that a new perspective is crystallizing regarding our understanding of fraud in research. One can find many texts focused on fraud in research. Many of these either describe cases or express concern or emotional anxiety, or they preach about the potential dangers that lurk behind such deviant practices. Some even offer remedies and explanations. While much of the literature focuses on more specific cases or practices, this book offers an "organizational field" system approach to research fraud. In doing so, it identifies the major stakeholders in the field and observes collaborative efforts between them to deal with research fraud. Research fraud is no longer an issue that is dealt with (or avoided) by universities, but it is of interest to research funding agencies, scientific and professional associations, journal editors, journalists, colleagues and collaborators, students and mentors, as well as the more popular media. Furthermore, the data suggest that since the establishment of the Office of Research Integrity as a leading agency in the United States that deals with research ethics, we could observe an increase in numbers of documented suspected cases of research fraud for which there is publicly available information.

The book is based on a database of 748 U.S. and global cases on which there are public records (in English) on suspected fraud and on a file with 2,839 pages of data sources on the fraud cases detected. As we made progress, it became clear that the key word in the unfolding story of fraud in research is "complexity." This complexity is expressed in a number of key issues, and focused on such issues as: partial available information; difficulties in identifying the exact nature of the issues involved (especially in plagiarism); problems identifying the exact nature of reactions to fraud; how and when suspicions are raised; how is the reality of fraud established; and, later, finding the exact reasons for organizational responses. Moreover, many researchers are adept in developing interpretations, and this wonderful quality was used in various cases to develop controversies, obfuscate, and in some cases make what appears to be a bona fide fraud look like something different. Along the way, it also became clear that it is not really possible to study fraud in research and avoid some serious deliberations about truth, deception, and moral boundaries. And indeed, discussing these topics provided us with numerous interesting debates and insights. Moreover, discussing fraud and research led us to notice that deviance and research can have two very different meanings: deviant researchers vs. deviant research (or science).

These complexities are integrated into the content and structure of this book. While the first three chapters are focused on the empirical reality of fraud within its organizational context, the following two chapters use the topic of this book to present, discuss, and expand more general topics of truth and untruth, of lies and fraud, and of fraudsters and control, and to develop a critical view of research norms.

MISCONDUCT

The term that is used frequently to refer to infractions in scientific research is "misconduct." Dictionary definitions of "misconduct" tend to compare the term to intentional wrongdoing, mismanagement, or improper behavior. What does this term cover in the context of science?

The range of behaviors under "scientific or research misconduct" is genuinely impressive. It includes such behaviors as plagiarism, falsifying reports, cheating, lying, false pretenses, inventing observations, cooking data, hired-hand research, abusing students, operating under conditions of conflicts of interests, false identifications, ethical issues, and perhaps a few more. One needs to hastily add that some of these behaviors in other areas such as the stock exchange, banking, or other private firms will probably result in a criminal police investigation. We can tell lies, of course, and as a matter of course we do. But once such lies are used to gain such benefits as rank, promotion, income, or prestige, then these lies acquire a different meaning. Depending on the severity of the deception, different actions can take place, from an administrative demand to rectify the falsification, to disciplinary court action, to police investigation.

Hence not every manipulated or falsified CV requires police intervention. Duping a granting fund to spend huge amounts of money on fabricated and falsified research may in fact require such intervention. Misconduct, in simple terms, refers to a variety of phenomena in which we have a research scientist that knowingly, deliberately, and intentionally violates the norms of scientific research.

We believe that—in terms of impact—two forms of fraud in research are more salient than others. The second is plagiarism, but the first is data fraud. While the second (and other similar research deviancies like hired-hand research) robs researchers of proper credit and certainly violates intellectual property rights, possibly causing serious personal frustration, pain, and damage, it does not change our understanding of the research topic, of the puzzle that was solved or the scientific body of knowledge. The impact of activities such as entering data that do not exist or eliminating other "interfering" data, however, is much more serious because it may change our body of knowledge and thus warp our understanding of potentially important aspects of our reality. Such fraud violates the most basic component of (and reason for) doing research.

DEVIANCE VS. MISTAKES

By deviance in scientific research we mean fraudulent acts committed consciously, deliberately, and intentionally by scientists. These acts—falsifying and fabricating data, fraud, plagiarism, cooking or massaging data—run counter to the scientific ethos. One has to distinguish between these intentional fraudulent acts and simple, honest mistakes.

If research scientists do their utmost to research a specific problem within existing conditions and methodologies and make honest, innocent mistakes, this certainly does not constitute an intentional act of deviance. One example involves the American astronomer Percival Lowell's report that he was sure he saw canals on Mars. Another interesting—however complicated—example involves Mendel's work, reported by sociologist H. Zuckerman (1977 105–6). The statistician R. A. Fisher claimed in 1936 that Mendel, considered to be the father of genetics, could not have achieved the results he reported because the agreement with theoretical expectations was far too close. Did Mendel commit a fraud? Not according to Dunn and Wright who, in 1965 and 1966 respectively, pointed out that Mendel's procedures have to be understood in the context of what was known about statistics and probability at the time he conducted his experiments. Dunn and Wright point out that although Mendel's results were biased, no fraudulent practice was involved. Although limited in generalizability, this case illustrates a potential complexity in distinguishing fraud in research vs. honest mistakes. It is possible that research utilizing contemporary knowledge and technology may appear suspicious when viewed later by researchers using more advanced knowledge and technology.

There are cases where researchers believe in things that are not verifiable by scientific observations, but these do not directly affect the contributions of the researchers. Such cases do not reflect a "mistake" but do not constitute acts of deliberate fraud either. For example, William Herschel, the discoverer of the planet Uranus, thought that the sun was cold, dark, and habitable under its flaming atmosphere. Johannes Kepler was a firm believer in astrology. Many famous research scientists believed in spiritualism and a few Nobel Prize winners believe in "a hidden spiritual world" (Gliedman 1982). A famous source for mistakes in the social sciences is due to so-called "demand characteristics," especially in psychological experimental studies. Psychiatrist and psychologist Martin T. Orne (1969, 146) defined "demand characteristics" as "the cues . . . which communicate what is expected of (the subject) and what the experimenter hopes to find." A second similar problem concerns experimental biases that are created by the effect of the experimenter's expectations from the subject's performance, mostly in psychological or social psychological studies. It was found that in some cases, either visual or auditory cues from experimenters—in most cases not intentional—were received by experimental subjects and interpreted and understood by them (Rosenthal et al. 1979). It is interesting to examine how scientists who err explain their mistakes. Mulkay and Gilbert (1982, 181) pointed out that "[w]hereas correct belief is portrayed as exclusively a cognitive phenomenon, as arising unproblematically out of rational assessments of experimental evidence, incorrect belief is viewed as involving the intrusion of distorting social and psychological factors into the cognitive domain" and that "[a]symmetrical accounting for error and for correct belief is a social device which reinforces the traditional conception of scientific rationality and which makes the community of scientists appear as the kind of community we, and they, recognize as scientific" (166).

Zuckerman (1977, 110–13) distinguished between what she called reputable and disreputable errors in science. By reputable errors she meant all the unintentional errors made by scientists, despite their attention to strict hypothesis formulation and methodologies. Disreputable errors, on the other hand, are errors such as not following the right methodologies or omitting important elements of experimental designs. That is, disreputable errors imply the intention not to follow procedures and are, in fact, deviances. When in 1953 Longmuir coined the term "pathological science," he had in mind all those scientific claims made due to both neglect of adequate methodologies and self-deception.[38]

The line between making an "honest" mistake and an intentional one is not always clear.[39] Gardner (1952, 1981), Zuckerman (1977), and this book in different parts document cases of research scientists committing or abetting fraud because of preconceived ideas or because of internal (scientific) or external (cultural) factors. Historian of science Steven Shapin (1982) attempted to describe some of the external (political, cultural) factors contributing to the development of various scientific ideas. Cornell University professor W. B. Provine (1973) studied the inferences drawn by biologists from their research on the genetics of racial mixture and found that the morals of the scientists changed their beliefs without any actual change in their knowledge. Sociologist Gwynn Nettler (1972) noted that "frequently scientific hypotheses are accepted or rejected, not just because of the evidence available, but also because of a moral pressure to interpret the evidence in a congenial way" (167). The rise and fall of "Project Camelot" (Horowitz 1974[40]) is another example. The cultural matrix in which science operates can thus lead to deviant acts, of which the Lysenko case (discussed in chapter 4) is just one glaring example.

Making a distinction between honest mistakes and deliberately fraudulent practices brings up another issue: sloppiness and ignorance. It may be that sloppy or ignorant research involves a higher cost than does fraudulent research. Honest mistakes may also have a high cost. This, obviously, is an interesting issue mostly because one can claim that sloppiness or ignorance are not that much different from intentionally lying. We feel, however, that a serious and significant difference exists here. Intentionally and deliberately fraudulent research practices involving lies and deceptions present a qualitatively different—and sinister—form of conduct than making mistakes, being sloppy in one's research, or even being ignorant. Obviously, ignorant, sloppy, mistaken, and fraudulent research should be kept to minimum, but they are not the same behaviors.

FORMS OF DEVIANCE

Deviance in scientific research can assume several forms.[41] Philosopher Charles Babbage (1975)[42] identified three forms of deviance: forging, trimming, and cooking. Forging (or falsification) means the recording of observations that were never made (e.g., the Summerlin, Kammerer, and Burt cases). Babbage stated that forging is motivated by the hunger for scientific reputation. Trimming means that the researcher manipulates data so that the results either support or at least do not contradict the theory.[43] Such cases are, naturally, hard to detect. Cooking means that the researcher reports selectively (see Berkowitz's 1971 study) and suppresses data. Babbage added hoaxing as a form of deviance, but for Babbage hoaxing was more like playing a practical joke (e.g., the case of the French chemist, Moissan, discussed in chapter four, who in 1893 thought that he had "manufactured" artificial diamonds while it is possible that this was a hoax perpetrated by one of his students). Babbage felt that these practical jokes become true deviance only if they are repeated and taken seriously. Unfortunately, hoaxing can become a serious matter. The whole field of UFO study fell into disrepute partly because of the large number of hoaxes perpetrated. The Piltdown hoax,[44] various hoaxed inventions, and other hoaxes (e.g., MacDougall 1940) do not lend much support to Babbage's comparison of hoaxes to practical jokes. Many of these hoaxes were definitely not meant to be harmless. For example, between the years 1976 and 1979, Italian inventor Aldo Bonassoli bilked the French government out of $150 million for an "oil sniffer" device that proved to be a hoax.[45] A more recent scam is the bogus bomb detectors sold to about twenty countries (including Iraq and Pakistan) by a British businessman who ended up with a ten-year prison sentence.[46]

Zuckerman (1977) added three forms of deviance: fabrication, manipulation of data, and data suppression, which are similar to Babbage's. Making distinctions between two categories used often by researchers—"falsification" and "fabrication"—is, however, not an easy undertaking and distinguishing between them can sometimes be difficult.

Canadian sustainability expert Beth Savan's work (1988, 146–47) detailed thirteen more specific patterns of deceit in science, among them invention of entire experiments, invention of data, alteration of data, suppression of inconvenient data, suppression of unwelcome projects, designing experiments that do not test any hypothesis, adoption of invalid or dubious assumptions, interpreting experimental results in a way that supports a predetermined direction or that supports a specific theory, plagiarism for profit or credit. Our database has plenty of illustrations for all these patterns.

Based on characterizations by the American Office of Science and Technology Policy, linguist and philologist Wilfried Decco (2002, 37) states that fabrication is "making up data or results and recording or reporting them" and that falsification is "manipulating research materials, equipment, or processes, or changing or omitting data or results such that the research is not accurately represented in the research record."

One interesting case took place in 2005 and involved "Professor Reiner Protsch von Zieten" who "was a professor of anthropology at Frankfurt University for 30 years before he was forced to resign in disgrace. A university committee revealed that throughout his career Protsch had plagiarized the work of colleagues and had systematically falsified dates on numerous "stone age" fossils, including a skull fragment—dubbed Hahnhöfersand Man—that allegedly linked humans and Neanderthals."[47]

Another case took place in 2006 and involved Jon Sudbo, who was a "doctoral student and faculty member, University of Oslo, and former physician in the Department of Medical Oncology and Radiotherapy, NRH."[48] It was found that he "engaged in scientific misconduct by reporting fabricated and/or falsified research in [a] grant application . . . submitted to the National Cancer Institute (NCI), National Institutes of Health (NIH), and its first-year progress report." It was found "that Dr. Sudbo engaged in scientific misconduct by falsifying and fabricating research" about oral cancer. Dr. Sudbo reported fabricated/falsified results for the effects of lesion ploidy upon survival in patients with oral premalignant lesions. In the Preliminary Data section of the grant application, Dr. Sudbo reported several events intended to demonstrate his experience in the research field that the Investigation Commission stated "appear as pure fiction." Also, in the first yearly progress report for the funded study, Dr. Sudbo falsified the number of patients that had been screened for admission to the study.[49]

In addition to three publications for which Dr. Sudbo admitted falsifying and/or fabricating data, the Investigation Commission found at least twelve other publications that warranted retraction because they could not be considered valid." These "publications address the same general research area as that addressed in the grant application and demonstrate a pervasive pattern of falsification/fabrication in research reporting on the part of Dr. Sudbo. The falsified/fabricated data presented in the grant application purport to demonstrate the feasibility of preventing cancer in a high risk population with nontoxic oral agents."[50]

Other forms of deviance in scientific research include plagiarism,[51] "stealing" of ideas, taking unjustified credit for a discovery (or work), and using students' work without giving them proper credit. Brian Martin's work, for example, places emphasis on exploitation of students, research assistants, and spouses, most commonly by claiming undue credit for their work. This is an example of what he refers to as "institutionalized plagiarism" (Martin 1994, 2013).

One other form of deviance research scientists may commit falls into the classification of unethical behavior: for example, experimenting with human subjects without their informed consent. Issues of informed consent and research integrity have become quite complex and this book is probably not the place to delve too deeply into them. The issue of "research integrity" is dealt with elsewhere in this book, mostly as a bypass to avoid using terminologies of fraud. But when subjects are a focus of research without their knowledge or consent and when their rights are violated, then that is a problem of deviance. Both authors of this book are members of the American Sociological Association (ASA), and the ASA's code of ethics makes it very clear that: "Sociologists do not involve a human being as a subject in research without the informed consent of the subject or the subject's legally authorized representative."[52] Thus, while exceptions may exist, in principle doing research by deceiving subjects is a form of fraud.

Social psychologist Hans Toch has pointed out that the institution of science is rife with competition and rivalry and that "[i]t is a tamed jungle, with 'ethics' marking the delicate intersection of its veneer and its core. We therefore invoke 'ethics' to keep the mess under control, but invoke it sparingly to assure continuance of the game" (191–92). Zuckerman (1977, 119–22) also put under the "unethical" heading behavior such as not giving proper credit to authors. Nevertheless, she felt that ethical deviance in science was a limited rather than a general problem. In summary, ethical problems in science probably revolve around such axes as the relations between science and the public, the relations among scientists, sex discrimination in science, topics for research, and rights of subjects (see also D'Angelo 2012).

As noted earlier, we feel that at its core fraud in research has to do with the scientific work itself, which is research. Consequently, we focus on such fraudulent behavior as fabrication, falsification, misrepresentation, cooking or trimming data, and sometimes hoaxing. Added to these are plagiarism and ethical violations that have to do with relations between colleagues,[53] toward or by students or relevant organizations.

TWO | Fraud in Research: Frequency Patterns

In this chapter we will present and explore the meaning of some of the patterns depicted by the cases of research fraud in our study's database. Before delving into that, an explanation of the methodology we use is in order.

METHODOLOGICAL NOTES

Like so many financial institutions that have to face cases of fraud, universities and research institutes feel embarrassed and uncomfortable when they have to deal with suspicions of misconduct. Thus evasion, secrecy, inconclusiveness, debates, arguments, and explanations are a common pattern. Moreover, in some suspected cases of research fraud we witness various stakeholders try to obfuscate and fog the issues. Another common denominator is that accessing data about fraud in scientific research is a difficult, sometimes impossible, task.

Universities have traditionally been reluctant to disclose cases of what is typically termed as misconduct or scientific deviance, and many maintain that such cases are rare and unrepresentative. Moreover, instead of using a rhetoric of fraud, universities bypass the core issue—fraud—by resorting to a rhetoric of misconduct, integrity, or ethics. And yet while the police have specialized units that have the knowledge and expertise to investigate cases of fraud and deceit, universities do not have such units. In addition, no specialized training exists on how to deal with such cases, and decisions have typically been made on an ad hoc basis or by ad hoc committees. An enforceable universal protocol of how to investigate, judge, and process such cases does not exist.[1] Thus many cases are shrouded in mystery, and information about them can be easily referred to as "dirty and hidden" (Ben-Yehuda 1990a). In fact, UK universities seem to have been so successful in their resistance that an agency like the Office of Research Integrity (ORI) in the United States was prevented from being established. The universities claimed that such an agency constitutes an incursion on institutional autonomy. Instead, the UKRIO[2] is a consultancy service that gets involved on request only, and as compared to the ORI is an inadequate website as far as recording and understanding fraud in research in the UK.

Plagiarism may present a particularly difficult challenge because sometimes establishing deliberate plagiarism is not simple. Issues concerning the nature of giving proper credit, being influenced by ideas vs. "stealing" ideas, simultaneous discoveries, or interpretations may create a gray area where the intent to plagiarize is difficult to establish. While today there is computer text-matching software that can—under certain conditions—identify similar texts and thus raise the suspicion of plagiarism, human examination is always required to actually decide if plagiarism is involved (see also Biagioli 2012). If quotation marks or proper references are provided then matching texts may not necessarily constitute a case of plagiarism. Nevertheless, with so much of the scientific research floating in cyberspace, such text-matching programs can have an important role in identifying plagiarism. Yet even with this technology, establishing bona fide cases of plagiarism may not be simple. One recent illustration involves Mike Sutton, a criminologist, who relied on text-mining software to claim in an e-book he published in 2014 (*Nullius in Verba: Darwin's Greatest Secret*) that Charles Darwin and Alfred Wallace plagiarized the theory of natural selection from Scottish naturalist Patrick Matthew. This claim—as could be expected—created much turmoil and remained controversial.[3]

A recent study conducted by Honig and Bedi (2012) used such software to examine 279 papers submitted for presentation at the International Management Division of the 2009 Academy of Management conference.[4] Their findings were alarming: 25 percent of the sample showed some amount of plagiarism, but more than 13 percent of the submitted papers exhibited significant plagiarism. The extent to which the complexities involved in suspicions of plagiarism can go are perhaps illustrated by the case of Kirkwood Community College's president Mick Starcevich. He used material from the *New York Times,* without attribution, in a speech he gave. After he admitted it and apologized, the Board of the college expressed its disappointment.[5]

Cases of fraud in research tend to belong to "the dark side of organizations" (Vaughan 1999) and are kept hidden for various reasons. We can thus only discuss cases that were detected or suspected, but we can only guess how many more such cases are taking place in any research discipline area, institution, region, country, or culture. Moreover, even when information is available, it is not always complete. Obviously this situation requires using as many sources as possible, be as careful as possible with the data, and try to cross-reference data whenever possible. This constraint seriously limits our ability make overarching generalizations that are both valid and reliable. Throughout the book, we state and restate, at times possibly to the point of fatigue, that our findings are limited to what we could detect and detail from the various sources we have painstakingly searched. Thus the reader should be aware that our findings are about what could be found and not about what there really is in terms of actions taken by individual researchers or by their organizations if they did not become public information. We are pleased, however, that we can claim that our database is as wide and inclusive as possible for 2010, and that we have made every effort to detect all the cases we possibly could.

A few words are in order about the "dark figures" in studies that try to research deviance in and between organizations. The sociological or criminological literature on deviance has discussed the problems of unknown or unobserved data. It is perhaps not surprising that there was little or no documentation of deviance in science and that the available information is incomplete and lacks many parameters. Any respectable criminology textbook differentiates between visible and hidden levels of crime in a society. Criminological estimates for various types of crimes state that significant amounts and types of crimes are not reported officially.[6] It is estimated, for example, that there are about twelve instances of unreported fraud for every reported case.[7] This state of affairs exists even though there are large law enforcement agencies whose job and interest it is to record deviant and criminal activity. Unrecorded deviance and claims that deviance is much more prevalent than what can be estimated by official reporting (the "iceberg theory") is not unique to scientific research. The term "dark figure" indeed refers to all the deviant cases that are unknown because they are unreported or hidden. Sociologists Gary Marx (1984) and Ben-Yehuda (1990a) referred to such data as "dirty and hidden." Indeed, our experience is that searching for data on fraud in research involves searching for dirty and hidden data. In an interesting way a paradox may exist here. Professional obligations seem to require that such data not be hidden. Researchers enjoy and protect their rights and privileges, but their professional obligations do not stop there. One obligation is clearly self-policing. Prima facie, preventing exposure of research frauds and deceit stands counter to professional obligations. But individual deviant researchers—like other deviants—do not have an interest in exposing their fraudulent behavior in public. Organizations where fraud occurred also do not have a strong interest in exposing all the knowledge they have about fraud cases. Even ORI's exposure of suspected cases does not provide full details.

Between the detected and confirmed cases and those that will never be found, there is a wide range of possible cases. These may include cases of suspected research fraud that were publicized by a university or a research funding agency but were never investigated by any formal authority. There are many reasons why an institutional stakeholder opts to avoid any formal investigation. It may be an effort to reduce the harmful publicity to the institutions from such a case. It may reflect some intra-institutional arrangements that are handled with secrecy with no information reaching the public. There can be cases where there are formal investigations, but these can take a very long time to resolve, and at the time we have looked into these cases, no conclusions have been reached.

Police around the world are collecting crime information while facing similar challenges. Skogan (1984) for example shows that police records and statistics may be incomplete or may contain mistakes, and obviously, not all recorded crime cases are confirmed when brought to trial.

In the organizational literature, there are also indications that even if cases of misconduct were detected, organizations are reluctant to open this information to other members within the organization, not to mention to actors outside of the organization. In addition, not all literature on misconduct or wrongdoing considers it abnormal behavior. For example, in a recent critical analysis of theories of misconduct in and by organizations, researcher of organizations Donald Palmer (2012) argues that many cases of organizational "wrongdoing" can be explained as a normal phenomenon rather than as abnormal behavior. They can be caused by such normal social or organizational factors as administrative system accounts, situational social influence accounts, accidental social control accounts, or social control accounts (2012, 4). In addition, sociologist Diane Vaughan (1998) claims that data in general on organizational misconduct and decision making is limited and is usually available only for retrospective accounts. This is a common research problem for studying organizational misconduct as there are no data on decisions made in real time in organizations. As Vaughan also claims, organizations have incentives to act in unethical and secret ways about this information. In her words (1998, 24), "Competition for scarce resources encourages research institutions to falsify data in order to win grants and prestige; universities to violate NCAA recruiting regulations in order to guarantee winning athletic teams; police departments to violate the law to make arrests that bring recognition and funding; political parties and governments to commit illegalities to secure national and international power. In response to competitive pressures emanating from the external environment, according to the amoral calculator model, individuals attempt to achieve organization goals through violative behavior." Kumar echoed this view in 2008, stating, "Increasing number of researchers has led to increase in competition for scarce resources. Researchers have often tried to take the shortest route to success which may involve performing fraudulent research" (211).

Thus, from a rational choice perspective, both organizations and individuals in organizations, operating under pressure and competition for such scarce resources as research grants,[8] promotions, and prizes have powerful incentives to commit misconduct and keep it secret. Indeed, one of the main results of Barber et al.'s 1973 research is that "researchers who did not do well in the scientific or in the local-institutional competition for rewards were more likely than others to be involved in studies with less favorable risks-benefits ratios for subjects" (p. x) and "[b]ecause . . . taking advantage of human subjects can sometimes enable researchers to get better data quicker, these underrewarded researchers more frequently become involved in studies with unfavorable risks-benefits ratios for the subjects" (83).

In writing this chapter, we share the frustration regarding incomplete data and partial information on cases. Thus all our cases are of suspected misconduct. While some of them were confirmed through a formal investigation, others have remained cases in which suspicions were raised by one or more sources. Some cases ended because no one continued the process of investigation, and in some cases, there is ongoing investigation but we do not have a final verdict.[9] Beyond the frustration with regard to the lack of clear results of investigated cases, we acknowledge the limitation in this study due to data partiality. In this chapter, we note the number of cases on which we found data on any specific variable, and then the distribution of cases in each of the categories for that variable.

With this in mind, we started this project with the aim of tracking all possible cases of research fraud that were publicized at any time anywhere in the world. How did we proceed in searching and creating our database?

The starting point was the website of the Office of Research Integrity (ORI) (https://ori.hhs.gov/), where we found descriptive data on 431 cases. Then we searched for additional information that was available on these cases in various other sources including Wikipedia,[10] newspapers, university websites, and articles in disciplinary scientific journals and websites.

After collecting several cases, we worked on compiling a database where all cases were coded on all possible variables we could measure. This was an interactive process in which we established variables and then attached the variables to related and relevant categories. Then we tried to code a few cases and then checked whether our definition of variables and the different categories needed revisions and adaptations. The decision on the final set of variables and categories we had defined was kept always ad hoc, as we had to allow the required flexibility needed to capture the wealth and the richness in all the cases we have collected. Thus, even at a later stage in the coding of the data, if there was evidence for a need to add a variable or a category to our set, we were happy to do so, even if it meant that we had to backcheck previously coded cases.

Weekly meetings with our research assistants usually lasted around two hours. These meetings were dedicated to discussing the new cases that were discovered and issues that came up during the coding of the data. We documented all the decisions made and made all efforts to be consistent in the way we treated the cases and coded the variables.

Data sources. Table 2.1 details the different data sources on which we relied in our study. Most of our initial data came from formal sources that documented cases of research fraud. Slightly more than 50 percent of the cases came from ORI (54.1 percent of the 748 cases in our database), followed by the National Science Foundation (NSF) (14.8 percent of the cases), and then, in descending order cases were found through books on research fraud, the Internet, the Danish Committee on Scientific Dishonesty, ORI international and *Science* and the *Scientist* journals, as well as the *New York Times* and even Israeli newspapers such as *Ha'aretz*.

Obviously, the sources of the cases we found are associated with the biases in our study. The reliance on accounts of cases that originated from ORI means that most of our cases are from medical-oriented research, since ORI is a unit within the National Institute of Health (NIH). Cases that were found through NSF were mainly in the life and natural sciences. Since we did not find research agencies that represented international research or European, Australian, South American, or Asian funding agencies, we are lacking representative information on these regions. An exception is Denmark, which established its own committee on scientific dishonesty that made cases of research fraud publicly known. As noted earlier, while the relevant UK website (UKRIO) focuses on issues of integrity and ethics, it is useless as far as reporting about actual cases or how they were processed.

Main variables. The focus of the study was on known cases of scientific or research fraud and the way they were detected and processed by their organizations. This empirical framework set the specification of the main variables we had coded. The main variables and the categories we used to code the cases are listed in the appendix.[11]

Qualitative data. For each of the cases in our study we searched for all known and available information. This created a textual database in which we have—for each recorded case—all the information that appeared in various sources such as the ORI, Wikipedia, newspapers, scientific journals, professional or scientific association webpages, and any blog or Internet listing. These sources provided us with a rich and descriptive textual database about the different characteristics of cases and enabled us to have a close and immediate examination of the phenomena we were studying. Just to give an idea, the size of this database is 2,836 Microsoft Word pages. In addition, many of the cases were constructed in separate PDF format files. Obviously, the level and depth of publicly known information varied among different cases.

PLAN OF THIS CHAPTER AND DATA PRESENTATION

After describing in detail our data constraints, the data sources, and the process through which our data were collected, we will now introduce all the variables and categories in our database. We mainly introduce the frequency distributions of the main variables in the database. It is important

Table 2.1. Sources of Cases

	Source	Frequency	Percent	Valid Percent	Cumulative Percent
Valid	ORI	405	54.1	54.5	54.5
	ORI International	26	3.5	3.5	58.0
	Books	97	13.0	13.1	71.1
	Internet	50	6.7	6.7	77.8
	NSF	111	14.8	14.9	92.7
	The Danish Committees on Scientific Dishonesty	27	3.6	3.6	96.4
	Science Journal	15	2.0	2.0	98.4
	The Scientist	12	1.6	1.6	100.0
	Total	743	99.3	100.0	
Missing	System	5	0.7		
Total		748	100.0		

to note here that since our data are based on secondary sources, we were not able to collect full information on all variables for each case. We will therefore indicate, for each variable, the number of cases for which we have valid data and their percentage of all the cases. These frequencies and the statistical analyses we have performed reveal an interesting and instructive profile of fraud in research as well as some valuable insights such as who suspected misconduct and how cases of suspected fraud were investigated. While examining some of these findings is important and instructive, our main aim is to explore how the aggregated picture of the distributions of the main variables can lead to a wide and valuable understanding of the main patterns of fraud in research.

Two additional comments need to be made here. The first has to do with the nature of the data. We re-emphasize that the data we collected reflect *known* cases. We have no illusions that these known cases reflect the entire universe of cases of fraud in research. Moreover, we do not have a strong and tangible base to estimate exactly the ratio between known to hidden cases. Nevertheless, one cannot—indeed should not—ignore known cases and the conclusions that can and should be drawn from these known cases. A crop of 748 cases is not a small amount, but caution needs to be exercised. The second comment has to do with normalization. Briefly stated, normalization refers to a process of adjusting values to a common scale. Such a process can be paricularly useful in cross-national comparisons. For example, in measuring homicide rates, one can measure the actual number of homicides per 100,000 citizens. The ratios that emerge in different countries, states, or cities can provide an instructive comparison. We invested a great deal of time and energy discussing this issue, trying to figure out if and how to normalize the data we collected. We eventually decided not to normalize because it is impossible to compare our findings to a common scale and because we have only known cases and they have their biases, as we have explained above. Rather, we take the database as one universe of content. That is, for example, if researcher X fabricated data in Michigan and researcher Y fabricated data in Ontario, we assume that in essence they both violated a similar research norm. Indeed, we were interested in such differentiations as gender, countries, years, institutions, and type of infractions. Normalizing these data, however, would pose a serious problem. For example, do we normalize by number of scientists in each country in a certain discipline in each year? Or by scientific published research in each discipline? Or by the number of higher education institutions in each country in each year? Or by the number of scholars per discipline? Or by the specific type of fraud? Figures for each of these possible normalizing procedures will definitely change over the years, places, or disciplines. One can easily realize why this is a genuinely thorny issue capable of giving one a decent headache. To make things even more complicated, there are areas where small numbers exist (e.g., year, country, some disciplines). Normalizing (assuming an easy access to the different potential bases of normalization exists) can have the potential to "break" the database into a large number of specific items and may create a situation where one "can't see the forest for the trees." Thus, at the present state of research, our choice was to stay with absolute numbers and what they imply, very carefully.

CHARACTERISTICS OF INDIVIDUALS OR GROUPS OF SUSPECTED MISCONDUCT

Source that publicly disclosed the case. We wanted to know who was the source of the information about the publicly disclosed case of fraud. We have valid information on the source that disclosed the case in only 126 cases (16.8 percent of the 748 cases in our database). This information reveals that in 61.9 percent of these cases the misconduct became public knowledge through a scientific or a professional journal; in 38.1 percent of the cases it was revealed by a newspaper or magazine, and in 5.6 percent of the cases it was publicized by an Internet site, a radio program, or a TV program.

Distribution by country and state. We inquired about the country or state in which the case of fraud was found to learn whether the detected cases are geographically or nationally concentrated. The cases represented in our database take place in a number of countries and states. Because our main source of information was the U.S. Office of Research Integrity, a majority of the cases in our database took place in the United States (602 cases, or 86.95 percent of the total). In addition, we have 24 cases from Denmark, 13 cases from the UK, 11 cases from Israel, 10 cases from China, 10 cases from Germany, 7 cases from India, and 6 cases from Canada. These are the main countries where cases of research fraud were detected and made public. Countries with fewer cases of research fraud are not listed due to very small distributions.

Timing of cases. The distribution of cases of fraud in research over the years is an important issue as learning about general fluctuations in detection of suspected cases is of significance. Figure 2.1 depicts the years in

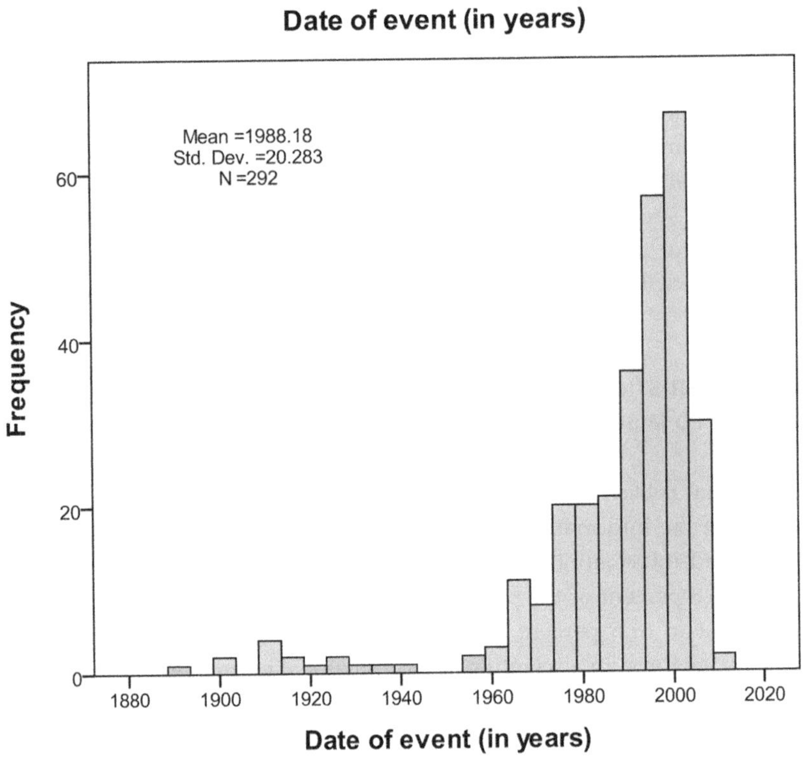

Fig. 2.1. Distribution of date of fraud in research event

Table 2.2. Distribution of the Main Countries or States where Misconduct was Reported

Main Countries	Cases (*N*)	Valid Percent
United States	602	86.95%
Denmark	24	3.2%
United Kingdom	13	1.8%
Israel	11	1.5%
Germany	10	1.3%
China	10	1.3%
India	7	0.9%
Canada	6	0.8%

which detected events of cases of fraud in research took place. We have valid data for only 292 cases.

It is not too difficult to see that there was a dramatic increase of reported cases beginning in the 1980s, which peaked between the 1990s and 2000 and then declined. This dynamic brings to the surface a conventional question. Can we disentangle whether this phenomenon is there because there were more cases of fraud or because organizations became more zealous about policing and publicizing fraud?[12] While a few possible explanations for this development may come to mind, as in other similar cases (e.g., police activity and amounts of crime), answering this question is not simple and the methodology we used is not conducive to providing a definite answer. Nevertheless, and given what we know, we will try to volunteer an educated guess. The years these phenomena took place were the years when the Office of Research Integrity was created and began to operate. Thus the explanation we find most plausible and persuasive is that as ORI's initial work increased awareness, detected more cases of research fraud, and started publicizing and exposing them to the public, the number of reported cases increased. But as the activity of this office continued, the number of reported cases declined. This fact may provide evidence about the institutional effect of the ORI on the organizational field and on the participating individuals within it. One possible explanation for this decline is that the activity of ORI and possibly other stakeholders increased awareness of the issue of fraud and stimulated the development of more and better preventive or control mechanisms. Obviously, one cannot avoid another more sinister possibility—that is, that those committing fraud began using more sophisticated methods of concealment and it thus became more difficult to detect them.

Some indirect potential support for the absolute increase comes from the growing number of retractions. "How many retractions? Some thirty plus years ago, there were very few retractions—on the order of 1 or 2 retractions *per* year. In the 1990s, the number rose to approximately 1 or 2 a month (i.e., 10–20 *per* year). By the late 2000s it rose to almost 1 per day (around 300 *per* year), and in the last couple of years it has risen to approximately 3 retractions every 2 days or around 500 *per* year" (Katavić 2014). Indeed, not all retractions involve suspicion of fraud, but the numbers are staggering, and Katavić raises "the possibility that the number of authors ready to commit scientific misconduct or abuse publication ethics is on the rise."

The amount of time it takes for a case of fraud to be investigated from beginning to end is an interesting and important issue. Lengthy processes may indicate how extremely careful organizations are not to punish innocent people, or in contrast it may indicate how organizations are dragging their feet, hoping they will not be discredited. Lengthy organizational investigations may also indicate that the organization lacks the resources needed to conduct such investigations.

Examining the data we found shows that in 4.7 percent of the cases the investigation took less than a year; in 37 percent of the cases it took one to three years; and in 51.9 percent of the cases it took three to five years. In less than 6 percent of the cases for which we have information, the investigation took more than fourteen years. The median length of investigation of suspected cases of research fraud was between two and three years. This clearly shows that the investigation of suspected misconduct cases is usually long, which probably indicates a complex investigation process.

Individuals or Groups. We examined whether suspected cases of fraud were committed by an individual or a group. Our findings indicate that most cases of suspected fraud involved individual researchers: 651 cases (87 percent of all known cases). In only 13 percent of the cases did the fraud involve more than one person.

An example of the potentially complicated relationships between a suspect in research fraud and others with whom he collaborated follows.[13] The example also shows how the process of revealing cases of fraud in research and treating them is conducted by collaborative action between multiple stakeholders. These include researchers from different universities who work in related areas, editors of journals where the research was published, employing organizations, scientific and professional associations, and even the university that granted the PhD. The example is about an outstanding young physicist, Jan Hendrik Schön, who worked at Bell Labs. Schön was considered a bright scientist and had won two prizes for his work in physics and materials sciences. Schön published several papers in the journal *Nature* presenting breakthrough advances in the fields of superconductivity and molecular electronics. However, researchers at Princeton and Cornell, however, were unable to replicate his results and found anomalies in his data, and they began to question the accuracy of his results.

In May 2002, after hearing about the suspicions, Bell Labs set up a committee chaired by Malcolm Beasley of Stanford University to investigate Schön's work. The committee obtained information from all of Schön's coauthors and interviewed the three principal ones (Zhenan Bao, Bertram Batlogg, and Christian Kloc). The committee also examined electronic drafts of the disputed papers including processed numeric data they presented. The committee requested copies of the raw data but found that Schön had kept no laboratory notebooks. He claimed that due to the limited hard drive space he had on his computer his raw data files had been erased. In addition, he claimed that all of his experimental samples had been discarded or damaged beyond repair.

He was accused of falsifying and fabricating data in a series of high-profile papers on superconductivity and molecular electronics, and his coauthors were initially suspected as well. Jan Hendrik Schön was fired after the investigating committee found him guilty of "scientific misconduct" on sixteen of twenty-four charges. All his coauthors, who had contributed to the experiments and whose names appeared on several of the papers, were cleared of any misconduct. The committee found that Schön showed reckless disregard for the sanctity of data in science and claimed that Schön substituted whole data sets to represent different materials or devices. In addition, he did not keep proper lab records and had deleted all the original, raw electronic data files. The committee wrote, "Such practices are completely unacceptable and represent scientific misconduct."[14] Schön admitted that he made mistakes and expressed his deep regrets, although he insisted that all his publications were based on experimental observations.

The complexity associated with confirming team fraud is associated with the fact that although the three collaborators were found not guilty of any charges of misconduct, the committee noted that "in one case questions remain that the committee felt unqualified to resolve, given the absence of a broader consensus on the nature of the participants in collaborative research endeavors."[15] This scandal led to an expansion of the ethics guideline by the American Physical Society, a professional organization that represented the nation's forty thousand physicists. The council of the American Physical Society also said that "[s]cientific misconduct diminishes the vital trust that scientists have in each other and undermines public confidence'." As a result, they called for institutions to offer researchers more ethics training and urged them to adopt the research procedures published by the U.S. Office of Science and Technology Policy in 2000.

The professional association's initial guidelines for proper scientific conduct were published in 1987 and updated in 1991. Dr. James Tsang, an IBM physicist who headed a panel on public affairs, felt they may have been unclear to members and that there was a need to refresh members' awareness of the problem of scientific fraud. He further said, "We needed to point out what good practice was in handling allegations of misconduct."[16]

The need for changing the policies was also attributed to the changing nature and organization of doing research in physics, moving from single-authored papers to multiauthored papers and multipeople, at times interdisciplinary, projects. These changes required new guidelines to clarify coauthors' roles and duties.

Interestingly, the results of the investigation into Schön's fraud have spilled over onto another institutional context. The *New York Times* announced two years later that Konstanz University (in Germany), which had awarded Schön his doctoral degree, had revoked it as a result of the fraud at Bell Labs. They made this decision even though there was no evidence that Schön had used any unethical procedures in the research for which he received his degree. Yet "the university said it had a legal right to rescind a degree when the recipient behaved 'unworthily' of it."[17]

Like in other cases of deviants, issues of gender and age are interesting and potentially important variables to examine.

Gender. Of the total number of cases about which we have information, 78.3 percent (432 cases) were committed by males and 19.7 percent (109 cases) were committed by females. There were eleven cases where the suspected fraud was committed by a mixed group of men and women (2 percent of the total). Thus the gender bias our data reflect is not only clear but consistent with similar gender biases in other areas, for example in crime statistics. Females are significatly less represented in various categories of crime and deviance.

Age. We have the age of the person committing the fraud in only ninety-eight cases. This relatively small sample limits our ability to offer strong generalizable conclusions. Due to the interesting pattern found here, however, we nevertheless cautiously discuss these findings. The mean age is 43.37 and the standard deviation is 10.7. Figure 2.2 provides a more detailed view. It indicates that there are two salient peaks of ages; the first one is around the ages of 34–35, and the second is around the ages of 42–43. The first peak is probably associated with what is typically the early stage of an academic career (ages 34–36) and the second peak with a more advanced stage (ages 41–44). These stages are typically associated with pressure to produce high-quality research to advance one's career in the academic world.

Academic rank of suspect. The academic rank of the suspect is known in 508 cases. The highest group of researchers suspected of research fraud occupy advanced tenured academic ranks (28.2%), the second group is of

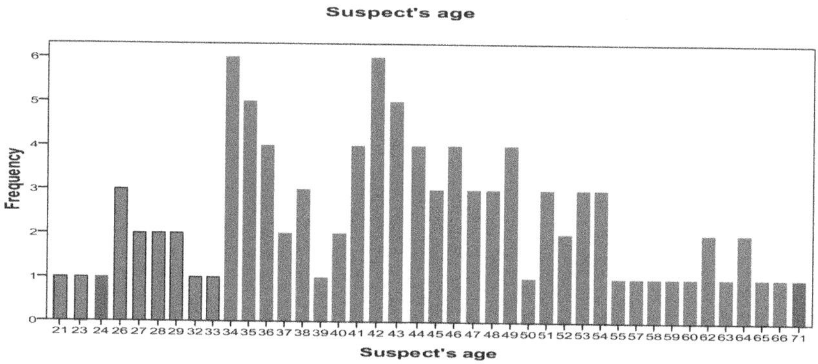

Fig. 2.2. Suspect Age by Frequency

postdocs (12.6%), then research staff (10.6%), and followed by untenured researchers (8.7%). The other small categories include research assistants, part-time research workers, nurses, administrative employees and others.

INSTITUTIONAL AND DISCIPLINARY CONTEXT OF FRAUD

Institution where suspects of fraud received their highest degree. In only 122 cases (16.3%) of suspected fraud cases we were able to obtain the relevant information. There is a great variety of academic institutions in which suspected individuals received their highest academic degrees, and only a few of them appear more than once. It may be of value to note that there were six cases from Harvard University and five from Harvard Medical School, and the second highest frequency in academic institution was UCLA, which had six cases.

Type of institutions where suspected cases of fraud occurred. As noted in the introduction, cases of fraud have occurred in different types of institutions including universities, hospitals, and research centers. Since a main source of our data is the Office of Research Integrity, which focuses on cases of fraud in research funded by the U.S. National Institute of Health (NIH), we acknowledge a possible bias toward finding more cases of research fraud in universities and hospitals and in the life and medical sciences, and less in other disciplines and in knowledge-intensive firms where applied research is conducted.

Figure 2.3 suggests that most known cases of fraud took place in universities (384 cases out of the 510 cases where such information was avail-

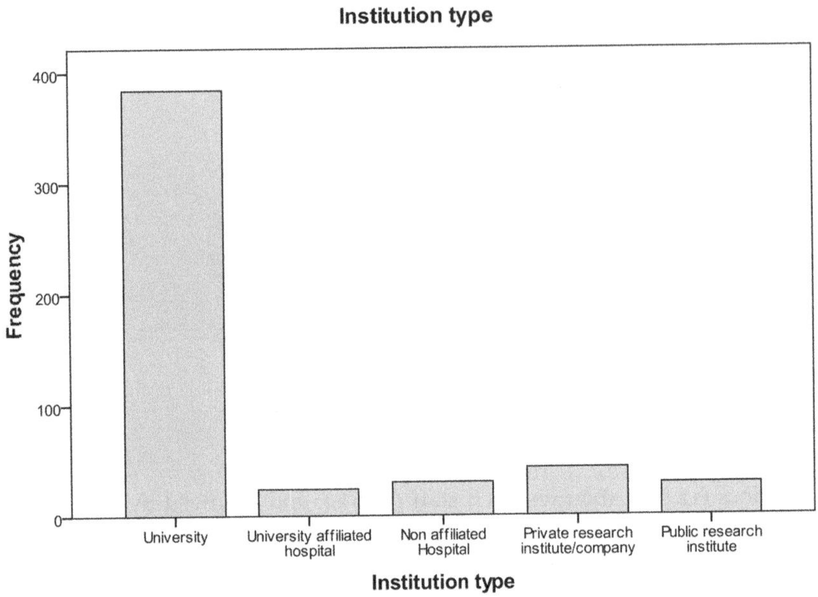

Fig. 2.3. Institution type by frequency

able, or 75.3 percent); 4.7 percent of the cases took place in university hospitals, and 5.9 percent of the cases in hospitals not affiliated with universities; 8.4 percent of the cases took place in private research institutions or companies.

Institution where suspected misconduct took place. Suspected misconduct cases took place in many academic institutions and hospitals. The list of universities or hospitals where alleged misconduct took place includes about four hundred different institutions. We were curious as to whether some institutions had more cases of suspected fraud than others and, if so, what we could learn from the frequencies. One might expect that more cases would be found in lower-ranking, less prestigious institutions, which might be assumed to have lower norms of scientific integrity. When we looked for the institutions where more than three cases of alleged misconduct took place, however, we were surprised to find that some of the most highly ranked universities were home to the greatest number of cases. Among these highly ranked institutions, including U.S. universities such as Harvard, UCLA, Northwestern, and Pittsburgh, the range of detected cases of misconduct was between six and eighteen.[18]

This finding could indicate several points: First, top universities may create an environment that requires a high level of quality research output and may thus pressure some researchers to falsify data or take other unethical shortcuts to publish their data more quickly. Second, it may be an indication that highly ranked universities have established effective monitoring policies that have led to a higher rate of detection of fraud. In addition, this finding may also indicate that when cases of fraud occur, highly ranked institutions—characterized by more openness—are more willing to expose such cases to the public. It is also possible that in cases where fraudulent data is published in prestigious journals (possibly by researchers from highly ranked universities), it would be difficult to cover up the fraud. Yet when problems are detected at an early stage they may be taken care of "in house" and not publicly announced.

Another factor may be that researchers working in such institutions are usually conducting research financed by prestigious funding agencies such as the National Institute of Health (NIH) and the National Science Foundation (NSF), and some cases may be detected by these funding agencies. Such detection may prevent universities from covering up the cases.

It would be interesting to conduct some intra-institutional analyses of the different cases of fraud that were found in these institutions, but such detailed analyses are beyond the scope of this book.

Disciplinary area. The general disciplinary area where the suspected fraud case was committed is presented in figure 2.4. We remind readers that our sample is biased because so many cases were found on the website of the NIH Office of Research Integrity, which, as its name suggests, funds research in the medical and life sciences. In addition, the information on the discipline of the suspected case was not always known.

The figures bellow reveal that 329 cases (55%) of suspected misconduct were conducted in disciplines associated with medicine, 185 cases (30.9%) with disciplines associated with the natural sciences, and 84 (14%) of the cases were in the disciplines of the social sciences and humanities.

Our findings indicate that within the medical sciences, the fields with the most suspected fraud activity were oncology (56 cases), immunology (36 cases), neurology (20 cases), psychiatry (20 cases), clinical trials (25 cases), and cardiology (12 cases).

Within the natural sciences, the most suspected fraud activity took place in biology/life sciences (96 cases) and chemistry (38 cases). In the social sciences and humanities, the most suspected fraud activity was found in psychology (36 cases) and history (15 cases).

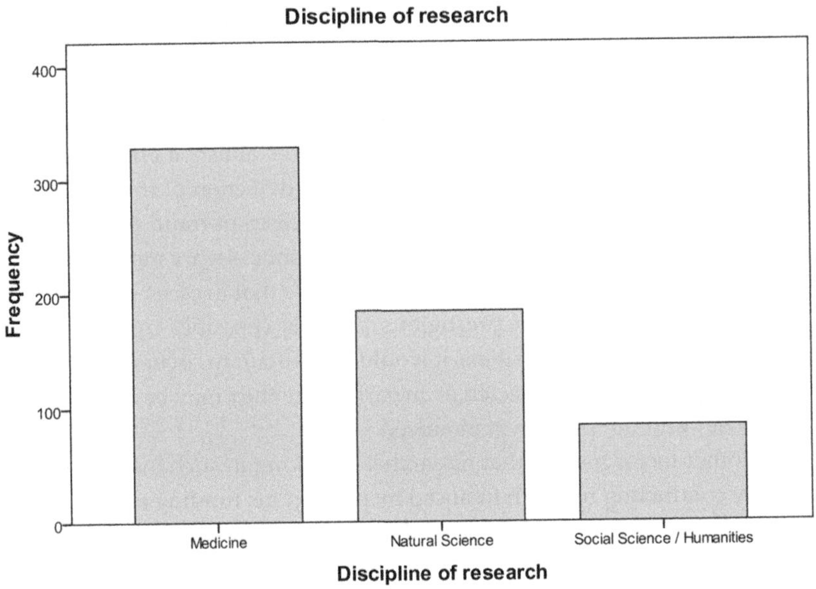

Fig. 2.4. Discipline by frequency

These distributions suggest an interesting potential hypothesis. They suggest that prominent researchers, many of whom are trained in top universities, go on to do medical research in leading universities. This may place the researchers in highly competitive environments where they are under pressure to be innovative and to produce positive and strong research results. Exposure to such environments may prompt some researches to think they can take fraudulent shortcuts and get away with it. Most research in medicine and the natural sciences, however, is conducted in research laboratories or within hospitals where the odds are high that any "irregularities" will be observed by other researchers. Thus, researchers doing fraudulent research in such environments run a high risk of being exposed.

Examining the very broad disciplinary distribution raises an interesting and potentially significant issue for future research. Obviously, organizational cultures and the form of rewards and incentives present a decent amount of variance between specific disciplines, within groups of disciplines, and in different organizational contexts. For just one example, pursuing research in biology in a university medical center or in a university faculty of natural sciences or in a liberal arts institution can exhibit different cultural ambiences. This could mean that detection and investigation of suspected cases of fraud could present very different challenges and complexities in diverse research contexts and disciplines. Consequently, an in-depth examination of patterns of suspected research fraud by specific disciplines with an emphasis on similar disciplines in different organizational settings may yield some interesting results. Such study, however, will require a different methodology and detailed information about a large number of cases of research fraud. Such detailed information on the scale required is not currently publicly available.

TYPE OF RESEARCH FRAUD

Table 2.3 shows the distribution of the various types of research fraud found in our database. In this table we are making a distinction between researchers who did not collect any data (fabrication), researchers who collected data but changed the numbers (falsification), and researchers who lied about how the data was collected and/or showed only the findings they wanted to highlight while concealing findings that contradicted their claims (misrepresentation). Plagiarism is defined as "[t]he use of ideas, concepts, words, or structures without appropriately acknowledging the source to benefit in a setting where originality is expected."[19] Violation of ethical rules means that the ethical rules of conducting scientific work were violated, and falsification of academic credentials (e.g., CVs) or publication status refers to intentional disinformation on degrees or credentials or on acceptance of papers for publications. Theft refers to taking materials from other scientists, and ghostwriting is when one person is writing research publications for another person without disclosing his or her identity or being involved in the study.

Table 2.3. Distribution of Types of Research Fraud

Type of Research Fraud	Cases (*N*)	Total Cases (%)
Data Falsification	381	51.0
Data Fabrication	288	38.7
Plagiarism	191	25.6
Data Misrepresentation	126	16.9
Violation of Ethical Rules	50	6.7
Falsification of Academic Credentials	37	5.0
Falsification of Publication Status	13	1.7
Theft	10	1.3
Ghost Writing	7	0.9

Note: Total number of cases on which we have data ranges from 745 to 747.

This table indicates that most cases of research fraud were associated with the manipulation of data, including falsification, fabrication, and misrepresentation.[20] The second category relates to published work and includes plagiarism or ghostwriting and falsification of publication status. Finally, the third category relates to general ethical issues in the way the research was conducted, such as the administration of consent forms or revealing conflicts of interest or theft of research materials.

HOW WAS THE SUSPECTED RESEARCH FRAUD FOUND?

Fraud, by definition, is committed in such a way as to avoid being detected. Yet the cases collected in this study were indeed detected, and most were subsequently investigated. From the organizational and deviance points of view, it is important to find out who uncovered the suspected fraud.

Funding agencies. Funding of research can expose the study to early inquiry by the funding agency. The 379 cases (70.4%) of alleged research fraud in our database for which we have information on the funding sources, were funded mainly by public funding agencies from U.S. national funding agencies such as NIH, NSF, and NIMH.[21] Interestingly, 162 (29.9%) suspected research fraud cases on which information is available were detected while applying for public funding.

Publication. Often the research fraud, of any nature, in our database had not been detected before the results were published in a scientific journal or the proceedings of a conference. Our data show that out of the 602 cases on which we have information, 64 percent of the cases where research fraud was suspected were associated with already published research, while in 36 percent of the cases the work was unpublished. This finding shows that while all scientific journals rely on scientist–peer reviewers to judge the quality of submitted articles, these reviewers played a relatively insignificant role in detecting fraudulent research. Suspected fraud was detected after the research passed the peer review and was published.

Colleagues. The data in table 2.4 below indicate that research collaborators detected the majority of the cases of fraud: 68 cases out of 321 on which we have data (21.2%). This may imply that local knowledge and interests may play a significant role in decisions of whistle-blowers to go public. But we need to view these actors within their context: that is, the organization.

And, as sociologists know all too well, individual actors' behavior tends to reflect cultural values and thus takes into account values, morals, and priorities of organizations. Facing the possibility of research fraud by a colleague always confronts researchers with painful issues of breach of trust and honesty, failed integrity, lack of fair use of resources, and violation of the basic value of scientific research: seeking a truth.

Generally, collaborators share many of the research-related details in a study and are in the best position to notice any unexpected or unfamiliar patterns presented or any questionable research procedure applied. The question resulting from this strong finding is "in what research context do we have this pattern or finding?"

Several inquiry directions are interesting here. First, are the odds of tracing fraud by collaborators or colleagues greater in certain disciplines than in others? What was the academic rank of the whistle-blowers and how does it relate to the rank of the offender?

With regard to the discipline, a cross-tabulation between the discipline and the cases of fraud that were detected by a collaborator reveals that most cases were in the medical sciences (30 out of 50 cases, 60%), followed by the natural sciences (19 out of 50 cases, 38%), while only one case detected by colleagues was in the social sciences. Of the 12 cases of fraud detected by the funding agency, 10 were in medicine, and of the 12 cases detected by the journal editor/reviewer 10 were also in medicine (83.3%). One has to remember that since the majority of our cases come from the ORI database, most cases detected are in the medical sciences.

The institutional affiliation also matters. Our data show that the highest rate of suspected research fraud detection by collaborators took place when the collaborators were affiliated with the same institution as the suspect and when the suspect was an advanced tenured academic. This happened in 43.2 percent of the cases (19 out of 44 cases in which we have the necessary information) and in 22 percent of the cases when the suspect was a postdoc (10 out of 44 cases). In 16 out of 24[22] cases (66%) the colleagues were from another institution and the suspect was an advanced tenured academic, and in seven of nine cases, when the discovery was made through an anonymous complaint, the suspect was an advanced tenured academic (77.8%). These two last findings may suggest that once a researcher is an advanced tenure academic, if there are suspicions of fraud, they mostly come from researchers in other institutions or through anonymous complaints.

Our data also show that scientific colleagues who are not collaborating on the suspect study act as important social control agents or whistle-blowers. The findings show that in 31 cases (9.7 percent of all cases detected), colleagues from the suspect's academic institution and in 43 cases (13.4 percent of all cases detected), colleagues from another academic institution were the initial social control agents or whistle-blowers. Overall, the first eight rows in table 2.4 point to those "in the loop," that is, those that are relatively close to the researchers in terms of their research interests. Those included in these first eight rows explain 70.2 percent of the variance.

A deeper examination of our database reveals a valuable pattern. We wanted to examine whether the type of research fraud is associated with the type of whistle-blower who detected the suspected research fraud. As described earlier, we divided the fraud cases into three main groups: those associated with data manipulation, those associated with the publications and those associated with ethical issues linked to the way the study was conducted. We wanted to know whether there are different kinds of whistle-blowers that are associated with these three kinds of research fraud.

The first category, cases of fraud that were associated with data manipulation, included all cases involving fabrication, falsification, or misrepresentation of data. Our data show that these cases were usually detected by a set of closely related actors including research collaborators (26.9%), an academic advisor (8.6%), a student (3.5%), colleagues from the same institution but not associated with the research (10.1%), and colleagues from other institutions (11%). Together this group uncovered 209 cases of the 336 (62%) cases in this category for which we have data on the whistle-blower.

Table 2.4. Main Whistle-Blowers or Social-Control Agents Who Uncovered Misconduct

Social Control Agent or Whistle-Blower	Cases (*N*)	Percent
Collaborators	68	21.2%
Colleagues from suspect's academic institution	31	9.7%
Colleagues from another academic institution	33	10.3%
Mentor	20	6.2%
Student	10	3.1%
Colleagues from abroad	10	3.1%
Victim of plagiarism or stolen data[a]	47	14.1%
Suspect's institution official	8	2.5%
Funding agency official	16	5.0%
Journal reviewer or editor	12	3.7%
Newspaper / Journalist	17	5.3%
Reviewer for the funding agency	17	5.3%
Anonymous	10	3.1%
Other	22	7%
Total	321	100%

[a]In 9 of these cases the victim was also a reviewer for a journal and in 5 cases the victim was a student.

When we analyzed the cases in the second category, where the fraud was related to the publications themselves, we refer to the cases that include plagiarism, falsification of publication status (e.g., claiming a paper was accepted for publication when it was not), or ghostwriting. For this second category, we found that the cases were mainly detected by the reviewers of the funding agency (13.3%) and by the victims of the offense including collaborators, students, and colleagues (mainly in cases of plagiarism) (42.9%). All these cases add up to 73 out of the 105 misconduct cases in this category (69.5%) on which we have data on the whistle-blowers.

Finally, the third category is based on cases associated with general ethical issues with regard to the scientific work. This category includes cases where there were violations of ethical rules in the conduct of the research and/or falsification of academic credentials or theft of materials. Different types of whistle-blowers detected the cases of fraud in this category. These included mainly research collaborators (10 cases), colleagues from the researcher's institution (7 cases) and from other institutions (6 cases), but also newspaper journalists (7 cases) and funding agency officials and reviewers (7 cases). Together, these add up to 37 of the 58 cases in this category (63.8%) for which we have data on the type of whistleblower.

When we examine the large picture that is depicted by these data (the 499 cases for which we have information on the source of the initial detection), we see a pattern of different types of whistle-blowers for different types of fraud. Cases of data manipulation are typically disclosed by someone from the perpetrator's inner circle, be it a collaborator, a student, or a scientific colleague from another institution or even another country who is familiar with the area of research and is capable of detecting the fraud. Obviously, such disclosures require that other researchers who are committed to doing honest research are willing to become whistle-blowers in such cases. Such willingness to disclose information on colleagues is not obvious. The authors are aware of many cases where collaborating researchers or colleagues who claimed they were aware of research misconduct by another researcher but were not willing to disclose this information so a formal investigation could be carried out.

On the other hand, cases of disclosing fraud clearly benefit from actors who have either experience or motivation in processing grievances. One example is research-funding agencies who suspect fraud, while another is victims of plagiarism. In such cases, where there is a strong motivation resulting from a perception of harm (on the agency or individual levels), it is expected that formal complaints will be placed. This typically takes place when the agents of the funding agency, or the victims, have witnessed acts of fraud and can provide the initial evidence required to persuade the powers that be to start an investigation. Once done, the process typically rolls on its way.

Cases of research fraud associated with the violation of different kinds of ethical rules (our third category) are typically detected by a wide range of whistle-blowers including close colleagues and collaborators as well as distant actors such as journalists and funding agencies. Since this category has the smallest number of cases in our database (only 58 cases), these findings are not too robust and should be taken with caution.

INSTITUTIONAL INVESTIGATION OF SUSPECTED RESEARCH FRAUD AND THE RESULTS OF THE INVESTIGATION

What happens after an institution finds out about the suspected misconduct? Our database reveals that in 93.3 percent of the cases in which there were suspicions (484 cases out of 519 on which we have information), the case was investigated by the institution. Further, in 358 of these cases, the investigation confirmed the suspicions and research fraud was determined. In 93 additional cases, irregularities regarding the conduct of suspected researchers were found, although not fraud. In our view, this is a meaningful finding. It indicates that once there were suspicions, in most cases (78 percent of the cases) the investigation was able to detect fraud or irregularities. To complete the picture, in only 8.5 percent of the cases did the investigation reveal no fraud or irregularities.

One possible conclusion is that the decision of an organization to investigate an allegation of fraud is probably based on what it considers strong initial evidence and/or good and reliable intelligence. Such information about the suspected case had probably been gathered prior to the formal investigation. Thus, universities that conduct investigations have a good screening for nonfrivolous accusations.

Table 2.5. Outcomes of Suspicious Cases of Misconduct

Investigation and Outcome	Cases (*N*)[a]	Valid Percent
Found misconduct	358	61.9
Found irregularities but no misconduct	93	16.1
Not enough evidence of misconduct	57	9.9
Didn't find misconduct or irregularities at all	49	8.5
There was no investigation	21	3.6
Total	578	100.0

[a]This table includes information only on cases on which we have information on the outcome of the investigation.

SANCTIONS AND PUNISHMENT FOR OFFENDERS ACCUSED OF MISCONDUCT

In an organizational field, where there are many stakeholders with an interest in preventing, detecting, and sanctioning research fraud, the range of sanctions varies depending on the sanctioning stakeholder (Leahey and Montgomery 2011). As a result, if the sanctioning organization is the employing university, the employment contract will govern the sanctions applied, which will range from a note in the researcher's personnel record to being fired from the university. In other cases, when the sanctioning organization is a research-funding agency, the sanctions can include preventing the researcher from getting future funding or taking the researcher to court. It stands to reason that when fraudulent research has been published, the publishing journal will retract the paper. Is that a practice adhered to by journals? Our database does not provide a persuasive answer to this question, but it is clear that a journal can ask for a retraction of a paper once it is found to involve any kind of data manipulation or plagiarism.

Due to the variety of our data sources we only have partial information for some of the cases in our database, so it is impossible to give a full description of the outcome of the investigation and the sanctions applied by the different organizations in every case. Our findings can, however, describe the range of sanctions and the degree to which our data provide evidence that they were applied.

We have information on the different sanctions applied to researchers found guilty of fraud. The most common outcome is a temporary exclusion of the offender from academic activities. This took place in 228 cases (we found this information on 30.5 percent of all the cases). Other temporary sanctions are also sometimes applied, including temporary prohibition from submitting proposals to external funding agencies (16 cases, or 2.1 percent of the cases with known outcomes) and temporary suspension (21 cases, or 2.8 percent of the cases on which we have information).

These temporary sanctions are the lightest that are applied to offenders; the harshest is when the offender is fired or pressured to resign. Our data shows that in 60 cases (we found this information on 8 percent of all the cases) the researcher was fired, and in 80 cases (we found this information on 10.7 percent of all the cases) the researcher chose to resign.

Between cases with light sanctions (e.g., temporary sanctions) and heavy sanctions (e.g., termination of the employment contract), there are "mid-range" sanctions. These include some restrictions on academic activities (15 cases, 2.0%), removal from some positions (15 cases, 2.0%), and some economic sanctions (15 cases, 2.0%).

If the offender is a researcher working in a university or research center, the organization can exert control over the offender's future research by monitoring it or inspecting the research results prior to their publication. Our database shows that this is not a frequent sanction; it was applied in only 29 cases in our database, or 3.9 percent of the cases where the outcome is known. When funding agencies were involved in dealing with cases of fraud, however, they demanded control over the offender's future research in 156 cases (20.9 percent of the cases with known outcome).

Here are some examples of this sort of sanction. The first case is about Jusan Yang, a former molecular biology researcher at the University of Iowa who falsified and fabricated data in 2005.[23] The investigation of the case was conducted by the Office of Public Health and Science, which had funded the research jointly with the University of Iowa Office of Research Integrity. The investigation of the case showed that although the falsified data were removed from the manuscript prior to publication, they were presented at two professional conferences. The decision made by the Office of Public Health and Science was that "if he [the researcher] participates in research funded by the Office of Public Health and Science, he will have to contend with a series of restrictions for the next five years." In addition, "any institution that allows [this researcher] to participate in future research must submit a plan to the public-health agency detailing how he will be supervised and how it will ensure the integrity of his contribution."

A second example is an immunologist who reported faked data.[24] The Office of Research Integrity reported that the researcher, who was a postdoc at the Medical College of Georgia, manipulated data in a study supported by funds from the National Institutes of Health and the National Institute of Allergy and Infectious Diseases. According to the report, *Nature Medicine* accepted the paper by the immunologist and his research team on May 3, 2006. After the publication, a researcher in the same area of immunology who didn't work directly with the authors approached the journal after identifying duplication in one of the figures. The chief editor of *Nature Medicine* e-mailed the researchers and asked them to go back to the raw data. Only then did the researcher, who was a part of the research team, admit that he had falsified the findings and the figures. He then resigned from his position. The journal retracted the paper in September 2006, and the university conducted a formal investigation of the incident and sent the investigation's final report to the ORI in February 2007.

Since leaving the Medical College of Georgia, this researcher has continued his immunology research at the University of Alabama, Birmingham, and has published studies in some prestigious journals such as the *Journal of Immunology*. As in other cases, though, the terms of his voluntary settlement agreement with the ORI stipulate that any organization that receives U.S. Public Health Service funding for research where this researcher conducts research in the next two years must submit a supervisory plan for the researcher's activities to the ORI. Moreover, his research team must also provide proof to the ORI that the research was monitored and reported accurately in any publication. The researcher himself was also prohibited from serving on Public Health Service advisory committees and acting as a peer reviewer.

All institutional sanctions given after an investigation of misconduct cases are listed in table 2.6.

Table 2.6. Outcome and Institutional Sanctions

Outcome and Institutional Sanctions	Cases (*N*)	Percent of the Known Cases
The offender was temporarily excluded from some academic activities (mainly advisory boards)	228	30.5%
The offender was temporarily debarred from federal funding	176	23.5%
The funding agency requested control over offender's future research	156	20.9%
The offender resigned	80	10.7%
The institution fired the offender	60	8.0%
The offender was reprimanded by the agency	56	7.5%
The agency asked the offender or his co-authors to retract or correct a publication or a grant application	46	6.1%
The institution asked the offender or his coauthors to retract or correct a publication or grant application	33	4.4%
The institution will have control over the offender's future research	29	3.9%
The offender was forced by the agency to have ethics counseling or training	22	2.9%
The institution temporarily suspended the subject	21	2.8%
The offender was permanently excluded from some academic activities	17	2.3%
Some restrictions on the offender's academic activity	16	2.1%
The offender was temporarily prohibited from submitting proposals for external funding	15	2.0%
The institution removed the offender from some of his positions	15	2.0%
The institution applied some economic sanctions	15	2.0%
Withdrawal of offender's academic degree or rank	14	1.9%
The sides reached a settlement	6	1.1%

Note: The numbers in this frequency table are not based on mutually exclusive categories. The institution can place several sanctions on an offender and these variables were all based on a yes/no answer for each case. Therefore, when an offender received several sanctions, they were coded as 'yes' on the different relevant categories. Thus, offering a total sum to this table will be misleading.

WHEN RESEARCH FRAUD CASES GO TO COURT

Although the investigation and sanctioning of research fraud cases are on an upward trajectory, our data shows that only 40 cases (7.5 percent of the cases) were brought to trial in a court. Table 2.7 provides the details on whether formal legal procedures took place and what the outcomes were. Following are three illustrative examples of cases brought to court.

Case 1

On May 18, 1995, a federal judge in Baltimore ordered the University of Alabama and four of its researchers to pay the government and a researcher a total of $1.9 million for stealing the researcher's work and claiming it as their own to obtain federal grants.[25] The researcher whose work had been stolen, Dr. Pamela A. Berge, was an epidemiologist working for the pharmaceutical company Pfizer. She had gone to court under the Federal False Claims Act, which is rarely used for cases of scientific fraud. The report in the *New York Times* indicated that:

> Usually, there is little recourse for scientists who believe they have been cheated by colleagues if their university decides not to pursue an investigation. In this case, the University of Alabama made two brief inquiries, but ultimately decided not to dig too deeply, Dr. Berge said.

The trial focused on grants given by the National Institute of Health to the university and four scientists. Dr. Berge claimed that they had used her data (collected by her in 1987) without giving proper credit; that they had made false claims, presenting her work as if it was their own; and that they had used her work not only to gain grants but also to publish it in a scientific journal without crediting her. The court ruled in favor of Dr. Berge.

The judge imposed the maximum punishment. The university was ordered to pay $1.66 million: $498,000 to Dr. Berge and the remainder to the federal government. In addition, the four researchers were ordered to pay collectively $50,000 in compensatory damages to Dr. Berge and $215,000 in punitive damages.

Case 2

On Feb. 18, 2005, *Ha'aretz* (an Israeli newspaper) published suspicions raised against a leading Israeli gynecologist and the chair of the Gynecology Department at Tel Aviv University. He had allegedly forged signatures of physicians for an experiment on a medical treatment and had forged experimental results that he claimed were conducted on forty-one women.[26]

The story was widely covered by Israeli newspapers, which wrote that he resigned from all his duties and was taken to court and placed under house arrest for eight days. It is interesting to note that the claims about his work came from the physicians whose signatures were forged, from the previous head of the department who claimed that no such experiment was conducted in the department, and the suspicions of a journal editor who rejected the paper because the experiment did not make any sense. Since the suspect was also the chair of the Helsinki committee on experiments involving humans (in three hospitals), the head of the national ethics committee also expressed his concern that such a case of fraud, especially committed by the head of a regional ethics committee, would be harmful to the medical profession. Furthermore, the dean of the faculty of medicine at Tel Aviv University announced that the falsified experiment also led to the cancellation of the offender's academic degrees. The dean said that they share the responsibility and that they should have inquired about the data, even if their part in the study was minimal. He harshly criticized the three other physicians whose names also appeared on the published falsified paper, but there is no reference to any charges made against them. Another report published in *Ha'aretz*[27] said that further inquiry found fifteen additional suspects.

Table 2.7. Distribution of Trial (Yes/No) and Results

Trial and Outcomes	Number of Known Cases	Percent of the Known Cases
There was no trial	489	92.4
There was a trial and subject was found guilty	26	4.9
There was a trial and subject was found innocent	6	1.1
There was a trial and part of the allegations were substantiated	2	0.4
Reached a settlement	6	1.1
Total	529	100.0

Case 3

On July 23, 1994, the *New York Times* reported[28] that two universities had agreed to pay the federal government $1.6 million to settle charges that they helped cover up scientific fraud by a faculty member. This was the first time that an 1863 law, the False Claims Act, was used to recover money in a case of research fraud, and it thus opened the door to more cases being brought under that act. The case concerned the work of Dr. John L. Ninnemann, an immunologist who worked at the University of Utah and then at the University of California at San Diego, who was accused of falsifying information on papers about research related to immunological substances. The universities were accused of knowing about this fraud but not doing anything about it.

The universities denied the accusations. They claimed that Dr. Ninnemann's research was sloppy but did not constitute scientific fraud. The investigation of the case began in 1983 when a research colleague from the University of Utah told university officials that he believed Dr. Ninnemann had made false statements in his research. The investigators at Utah and from the University of California, where Dr. Ninnemann worked from during 1984 to 1988, investigated the matter but decided not to accuse him of fraud. Yet later on, the Office of Research Integrity investigated the issue for several years and brought it to court in 1989. They asked the Justice Department to take over the case because "the universities had defrauded the Government by permitting false claims to be made in scientific papers and by making false statements to the Government." Ninnemann was charged in court with falsifying information in at least nine scientific papers in four scientific journals. The two universities received grants on the basis of Dr. Ninnemann's false statements in grant applications, and they were aware of the falsifications following their own investigation. They nevertheless chose to falsely assure the National Institutes of Health that Dr. Ninnemann had corrected his statements. Both universities were found liable for damages because of their awareness of the fraud and because they failed to properly monitor Dr. Ninnemann's work and statements to the government.

Of interest is the University of Utah's reaction to the claim that the universities shared responsibility for Ninnemann's fraud. "Dr. Richard Koehn, vice president for research at the University of Utah, said that when the case first came up the university 'was not prepared to handle it appropriately.' But he added that the government was nevertheless 'overzealous' in pursuing the case." University of Utah president Arthur K. Smith said that the government should not try to "place on universities an obligation to guarantee the correctness of their faculty's research conclusions," adding that "[t]his is an impossible and self-defeating approach."

According to the court settlement in this case, the University of Utah agreed to pay $950,000 and the University of California at San Diego agreed to pay $625,000. The settlement included an agreement by the universities to establish new procedures to handle cases involving scientific fraud and to correct deficiencies in their current systems. In addition, it was agreed that Dr. Ninnemann would not admit guilt but would agree to be excluded from eligibility from all federal grants and contracts for three years. He also agreed to submit letters of retraction of five scientific articles published between 1978 and 1984 and letters of correction for four more scientific articles published between 1985 and 1987.[29]

When cases of fraud were taken to court, a majority of the accused were found guilty or part of the accusations were substantiated. The punishments ranged from prison with or without fines or probation (10 of the cases in our database) to fines only (9 of the cases).[30] Appealing the court ruling is a very rare event and occurred in only 13 cases. In most cases the results of such appeals upheld the court's decision (7 cases), and only in 2 cases did such appeals change the original decision. One appeal[31] (published on October 8, 2008) was in a case brought against Pfizer Drug Company, which was suspected of manipulating the publication of scientific studies to bolster the use of its epilepsy-prevention drug Neurontin while suppressing research that did not support its use. An expert who reviewed the documents, Dr. Kay Dickersin of the Johns Hopkins Bloomberg School of Public Health, concluded that the Pfizer documents spelled out "a publication strategy meant to convince physicians of Neurontin's effectiveness and misrepresent or suppress negative findings."

A prison sentence is indeed a most extreme punishment for scientific fraud. In only five cases of fraud in our sample was the suspected researcher sent to prison; the prison time ranged from less than a year up to three years.

An example of such a case is the one involving Eric Poehlman,[32] which came to its end in 2005 after more than five years of investigation and trial. Poehlman, a professor from the Vermont College of Medicine in Burlington, pleaded guilty to lying on a federal grant application and admitted to fabricating more than a decade's worth of scientific data on obesity, menopause, and aging. This fraudulent scientific work took place while Poehlman was conducting clinical research as a tenured faculty member at the University of Vermont. He was found guilty of presenting fraudulent data in lectures and in published papers and using this data to obtain millions of dollars in federal grants from the National Institutes of Health. He was the first scientist to be tried on such charges and could have received a sentence of as much as five years in a federal prison. After more than five years of denying the charges against him, lying under oath in court, and trying to discredit his accusers, Poehlman finally admitted his guilt. He was sentenced to one year and one day in a federal prison, and his case became one of the most expensive cases of scientific fraud in U.S. history.

Another group of stakeholders in the organizational field related to fraud in research includes the professional/scientific societies to which the suspected researcher belongs. In our sample, such professional or scientific societies were involved in the processing of only 18 cases (2.4 percent of the sample). In many of these cases, the scientific society heard about the case from public sources and reacted as well, but in only five of these cases was the offender also sanctioned by the scientific society.

One of the main issues regarding fraud in research is that the suspected scientific work is often published in scientific journals before it is revealed to be fraudulent. The question then is what is done with the published work and how can the scientific community know that the published papers should not be trusted or cited. The most common practice is to retract or correct the published work. A retraction is a statement that withdraws, cancels or refutes a published research paper. Such action may be initiated by the editors of a journal, by the author(s) of the papers, or by their university or research center. By retraction, the researcher requests that the journal inform its readers that the published work should not be used. Corrections can be made when there are innocent errors in the published work. Our data indicate that in 90 cases (12 percent of the cases on which we know the outcome of the investigation) the deviant researcher asked the journal to retract or correct the papers that were found to be problematic. In 78 cases (10.4 percent of the cases in which we know the outcome of the investigation) this retraction of a published paper also involved an organization (such as the university), a journal, or scientific collaborators of the offender who requested the retraction. It is indeed instructive that Katavić's 2014 work indicates that the number of retractions has increased dramatically and *Retraction Watch* noted in 2015 that "there have been about 1,500 papers retracted overall since 2012, when we first reported on the phenomenon, faked reviews have been responsible for about 15% of all retractions in the past three years."[33]

As Fang and Casadevall pointed out, retractions are "essential for correcting the literature and maintaining trust in the scientific process" (2011, 3855). But retractions as a possible recourse for dealing with falsified or flawed papers as well as a preventive measure for further citing falsified findings is not a simple procedure. To begin with, retracting a publication requires that a justified request is made and accepted. Then the retraction may come a long time after the original publication (during which time the retracted work may have been cited by others) and readers may be unaware of it. While retraction is probably the most important duty (and perhaps effective act) a journal can perform (Sharp 1991), there is evidence that retracted papers continue to be cited (Garfield and Welljams-Dorof 1990; Pfeifer and Snodgrass 1990). Finding ways that will efficiently alert other scientists about retracted research papers is a main challenge to the scientific community. The challenge is even more complex and demanding because the scientific community also needs to prevent new research from citing retracted papers in order to avoid the continuous impact of fraudulent research on other scientific work. Montgomery and Oliver (2017) discuss the problems associated with continuous citations of studies that contain bad data (i.e., fabricated, falsified, or erroneous). They claim that current practices for handling retractions are inadequate, and they use the metaphor of a "virus" to illustrate how such studies can spread and contaminate the knowledge system while they are treated as "valid." They draw from public health models that aim to prevent the spread of biological viruses and compare three models for treating retracted papers. These models include the governance model of professional self-regulation, the public health governance model, and a triple-helix model that requires a collaboration between the industry, the university, and the state that can offer a holistic treatment of the retraction process.

Another interesting and relevant question is whether or not the offender admits to committing fraud in the case under investigation. Admitting or confessing wrongdoing is a major moral and legal issue. The data show that in 179 cases (51.6 percent of the cases on which we have information on this topic and 23.9 percent of the entire database) the offenders admitted being involved in fraud, while in 168 cases (48.4 percent of the cases on which we have information on this topic, and 22.5 percent of the whole sample) the offenders did not admit to committing fraud. These findings suggest that cases of fraud in research are typically characterized by a strong evidential basis and consequently offenders may find it difficult not to admit fraud. Another possibility is that admitting fraud may mean a more lenient sanction. It needs to be added that there have been cases where the discussions took so much time, or that the blurring was so intense, that at the end it became unclear who did what or why. Yet it is surprising that after lengthy investigations along with strong evidence or testimony, so many offenders deny wrongdoing. Cases like the famous noncase referred to as the David Baltimore affair (described earlier) or—at the other extreme—the case involving suspicions that V. J. Gupta[34] falsified data illustrate how lengthy, sometimes obfuscating, and at times unclear processes can be. Gupta, a paleontologist from India, was accused by colleagues in 1989 of falsifying numerous fossil discoveries in the Himalayas. Yet despite the strong evidence they provided that his findings were impossible, he kept denying wrongdoing.

We have further analyzed the cases where the accused admitted fraud to see if the admission might be linked to the severity of sanctions imposed on the researcher. While the findings do not seem to indicate a strong pattern, it appears that most admissions of research fraud were associated with relatively light sanctions. We need to note, however, that light sentences also constituted the highest frequency of sanctions in our database. Thus the exclusivity of the above association may be spurious. When the sanction was an exclusion from advisory boards, 72.3 percent of the accused researchers admitted fraud (99 out of 137), and when the sanction was temporary debarment, 63.1 percent of the accused researchers (70 out of 111) admitted fraud. On the other hand, when the researcher was fired from the organization, 51.2 percent of them (21 of 41) admitted fraud. Could this indicate that researchers tend to admit fraud when they get relatively lenient sanctions because they have committed relatively minor violations? Or could this mean that researchers admit fraud in return for relatively lenient sanctions? Unfortunately, we cannot answer these questions. A further analysis should examine the relationship between the sanctions and the admission of guilt in the context of the type of accusations made against the researcher. This, however, would bring us down to very small numbers from which it would be impossible to generalize.

While admitting fraud is one aspect of closure to cases of fraud, expressions of regret are really what is expected in the field of scientific work that is based on norms of research integrity and trust. Only in fewer than 40 cases (of all our cases) did the offender apologize in public (either in court or in other public venue) for committing research fraud.

ORGANIZATIONAL CHANGES AS A RESULT OF FRAUD

Cases of fraud found by a university, a journal, a professional or scientific community, or a research-funding agency can be expressed as a "shock" to the system. In later chapters, we will see how detection and investigation of research fraud cases can be associated with changes in practices, procedures, or policies that took place in some of the relevant organization. We will see how journals have revised their instructions to authors who are submitting papers to them, we will see how universities applied new procedures to assure strong socialization and exposure to the practices of research integrity, and we will see how the court system can generate the legal incentives to pressure universities to initiate mentoring and monitoring of their researchers. This is especially relevant to those organizations that were found to be aware of fraud yet did not apply any corrective procedures.

Nevertheless, our data show that in only 23 cases (3.1 percent of our database) is there evidence that organizational changes intended to prevent future research fraud followed the detected case of fraud, an impressively small number. Examples of organizational changes that followed the investigation of cases of fraud that we found in our data were changes in internal procedures, policies, or rules with regard to their researchers, students, or data management.

In one case, Beijing University developed new rules for dealing with cases of research fraud.[35] In another case, the Karolinska Institute in Sweden began to consider introducing a compulsory system of archiving researchers' raw data with the aim of making manipulation more difficult. In yet another case, a university is mulling the establishment of a permanent committee or office to address research fraud.

A review and re-evaluation of the existing policies is another outcome. One university began to review its policy regarding cases of fraud and started establishing and promulgating a policy on the retention of materials produced under an NSF grant.

Mentoring, educating, and offering ongoing training of new and old faculty members regarding scientific integrity is another organizational change we have observed in the data. Indeed, some universities have taken steps to ensure that new faculty members learn the ethical requisites of proposal writing and that their senior colleagues play a more active mentoring role.

This also happened with respect to students. Another university in our database increased its efforts to inform graduate students about issues related to common practices and fraud in science.

There are some cases where the pressure on the organization to apply changes as a result of fraud comes from the researchers themselves. For example, 120 biologists signed a petition letter written by Xin-Yuan Fu, a researcher at Indiana University, calling for a government office to start investigating research fraud seriously.[36] In this case the pressure to change came from the bottom up. In other cases, the organizational changes are made top-down. For example, because some cases of research fraud have been detected, the Education Ministry of China[37] has set up a special commission to study fraud.

The requirements regarding research integrity of researchers working on U.S. Food and Drug Administration–related clinical trials are explicit and demanding. After a fraud case is found, however, there seems to be a need for even tighter monitoring. In this context, The U.S.-based National Cancer Institute[38] set up a Clinical Trials Monitoring Branch, which will make both scheduled and surprise audits at study sites to keep tabs on researchers. This new step obviously implies that the trust that had permeated scientific work is no longer taken for granted and that an assumption of suspicion has been introduced into the system.

Knowledge-intensive firms that confront cases of fraud also need to review their policies and procedures with regard to research integrity. For example, our data indicate that at Bell Labs,[39] existing policies about documenting research and keeping computer data were reinforced after a case of fraud in which one of their researchers was involved.

The organizational changes can also take place on the level of procedures that are to be specifically applied in a single laboratory. For example, after experiencing a case of fraud in his laboratory, a professor might want to take steps to ensure higher quality control by personally inspecting the primary data accompanying each finding.

Establishing new educational campaigns promoting ethics in research is another change that can be adopted by a university after a fraud case. As a result of a case of fraud, for example, the Department of Sociology at Peking University has launched an educational campaign among faculty members to promote academic ethics.

Finally, there are some organizational changes that are enforced by external stakeholders following a case of fraud. For example, subsequent to a case of fraud by a researcher at the drug company Merck, which was also brought to trial,[40] there was a settlement that included a few requests: that Merck would no longer engage in ghostwriting of articles written for publication in the media (where certain drugs are promoted as if research has shown their effectiveness); that Merck would submit all television advertising campaigns to the FDA for review before releasing them; and that Merck will adhere to any FDA recommendations.

DETECTION AND HANDLING OF RESEARCH FRAUD CASES OVER TIME

The detection of fraud cases is not independent of context. We claim that norms and regulations within organizations or in the general organizational field have an impact on the interests of organizations and individuals in detecting such cases and bringing them to the attention of the relevant stakeholders. Moreover, there can be little doubt that media attention to cases of fraud play a role as well. The answer to the question of what makes a specific case known thus depends on a few factors: inter- and intra-university regulations, interested persons, various stakeholders outside universities, and particularly the media. An analysis of detection of the three main fraud categories over time can either show an arbitrary pattern, meaning that there is no clear pattern for the detection of these cases, or it can reveal a pattern of detection for which field-based explanations can be offered.

In the next few figures we show the distribution of detected cases of fraud for the three main categories of falsification, fabrication, and plagiarism overtime. Since our database includes 431 cases of fraud (out of 748) that were initially found through the publications of the Office of Research Integrity, we have presented the graphs for all cases and then separately for the cases that appear in ORI.

The patterns found in these figures show clearly that the initial peak of detected cases of fraud was found in the years 1995–1996. This pattern was found both when the data came only from the ORI records and when it came from all the cases in our database. Another peak is at 2002–2004, although with some differences between the graphs.

The Office of Research Integrity (ORI) is the leading agency that was established to deal with scientific fraud in the United States. From the ORI website we can see how some time points are of significant historical value.[41] This is some of the text from the ORI website:

> In May 1992, OSI and OSIR were consolidated into the Office of Research Integrity (ORI) in the OASH. Later that year, HHS [Health and Human Services] established a hearing opportunity before the Research Integrity Adjudications Panel of the Departmental Appeals Board, HHS, for all scientists formally charged with research misconduct. In January 1993, the ORI published its first quarterly ORI Newsletter. In June 1993, the process of removing responsibility for

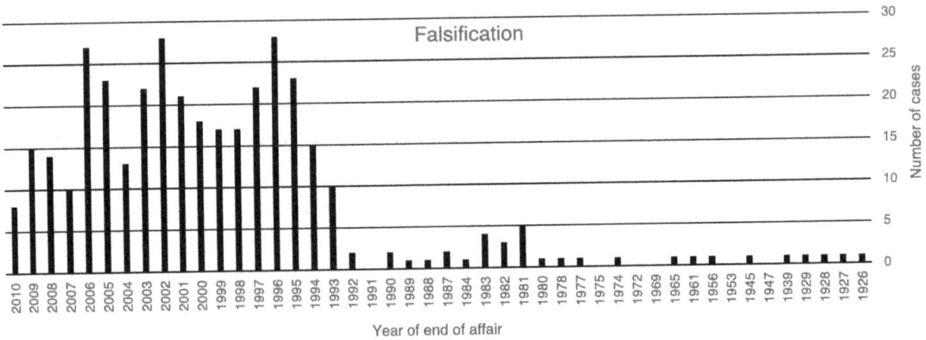

Fig. 2.5. Falsification: End of affair by count (all cases)

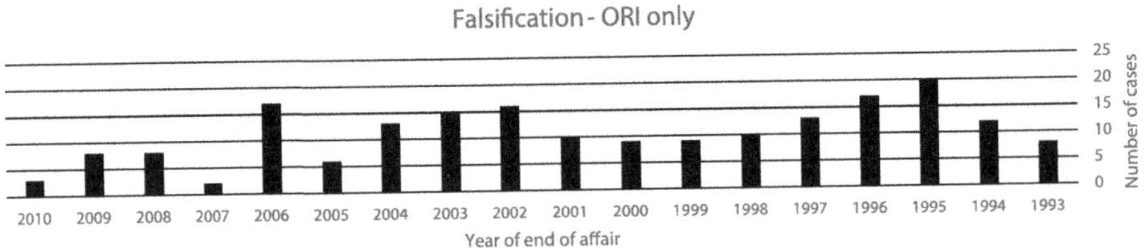

Fig. 2.6. Falsification, only ORI: End of affair by count

handling allegations of research misconduct from the funding agencies was completed when President Clinton signed the NIH Revitalization Act of 1993. This Act established the ORI as an independent entity within HHS and replaced the term "scientific misconduct" with "research misconduct". . . . ORI began an intramural research program in 1993 commissioning a study of the consequences of whistleblowing for whistleblowers that was completed in 1995. The first ORI Annual Report covered CY 1993 and was published in September 1994. The ORI web site was initiated in 1995. In 1996, the Secretary convened the HHS Review Group on Research Misconduct and Research Integrity (pdf) to examine the system under which the HHS handles allegations of research misconduct which included the issues addressed by the Ryan Commission. In 1997, ORI began its conference and workshop program by collaborating with an institution, the University of Florida, for the first time in organizing an event.

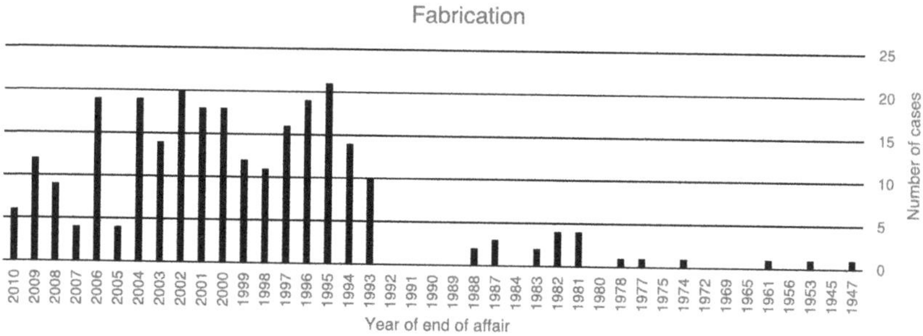

Fig. 2.7. Fabrication: End of affair by count (all cases)

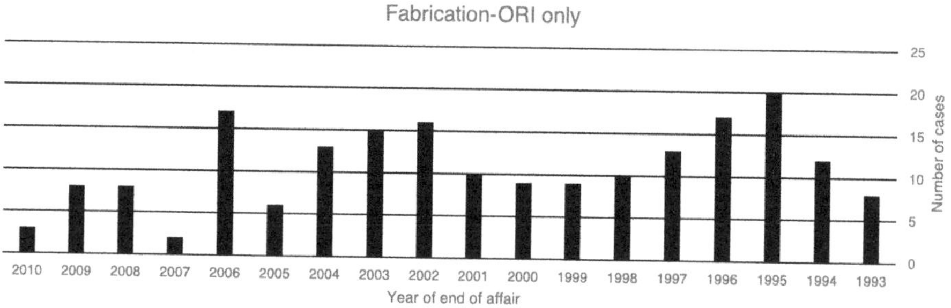

Fig. 2.8. Fabrication, only ORI: End of affair by count

If the official agency started in 1993 and the formal activities started in 1994–96, the number of investigations of research fraud seems to have had an initial peak impact in 1995–96. This increase is probably associated with the agency's activities. Later on, we can see that—despite some fluctuations—keeping a high level of detection and publicizing information by ORI probably accounts for the relatively high number of research fraud cases detected after the initial peak. The series of figures (2.5 through 2.10) show the patterns of detected suspicious cases over time based on the date of the end of the investigation for falsification, fabrication, and plagiarism. Each graph appears twice for each type of fraud, first with all cases we have information on and then for the cases from ORI only. These graphs depict the changes in the prevalence of each type of the main forms of fraud mentioned above over a long period of time. As noted, because ORI data plays a major part, we separated its data for each category so that readers could see how the introduction of the ORI led to a significant increase in the number of detected cases of research fraud.

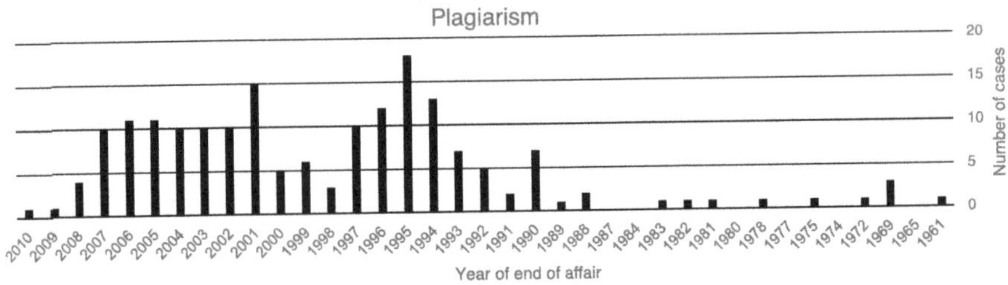

Fig. 2.9. Plagiarism: End of affair by count (all cases)

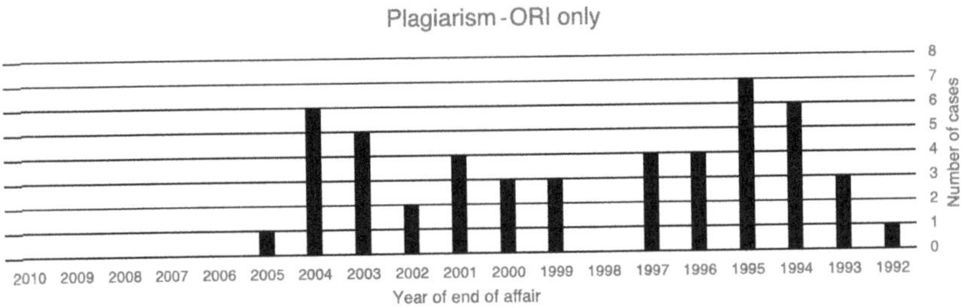

Fig. 2.10. Plagiarism, only ORI: End of affair by count

OFFICE OF RESEARCH INTEGRITY (ORI)

Before ending this chapter, a few words about ORI's data vs. the data bank we created are in order. ORI has published annual reports about its activities regarding fraud since 1993.[42] In the past few years these reports have been thick volumes reflecting the various activities in which ORI is involved. Among other things, ORI provides essential information on how to conduct an inquiry in cases of suspected fraud, provides resources for education related to ethical scientific conduct, gets involved in cases where public money is involved, and provides researchers and interested parties data about cases of fraud year by year. Nevertheless, our goal was not to follow ORI's statistics year by year but rather to compile one general database for all cases—both U.S. and global—and try to distill from it distinct patterns. Consequently, rather than focus on the relevant public arena in the United States exclusively, we examined all known cases, globally. We thus focused on some of the dynamics involved in a general longitudinal view and on some specific trends.[43]

Looking at some of ORI's statistics for the more recent 2008–2011 time period confirms similar observations from our dataset. There are many more suspicions (or allegations in ORI's terminology) of fraud than there were cases in which an inquiry was conducted. For example, out of 240 potential cases in 2011, only 44 cases were actually opened for inquiry. In 2010 only 28 new cases were opened while 155 allegations were registered. In 2009 out of 179 allegations, 9 became active and 49 were still assessed in detail. In 2008, out of 201 allegations, 17 new cases were opened. Clearly out of hundreds of allegations per year (an average of 198 for 1992–2007 and 271 for 2004–2006) only a few dozen cases triggered the launching of an inquiry. The road ORI seems to follow is to go from allegations to preinquiry, and from there to inquiry, and finally to an open case status. Each of these bureaucratic stages has its own statistics, which illustrates the complexity involved in investigating cases of research fraud. Our data show that less than half of the cases ever reached full investigation status that could confirm research fraud. This could be taken as an indication of the complexity of collecting valid evidence. This organizational process generates inter- and intra-organizational pressures concerning the suspected research fraud. In the next section, we will show and discuss the distributions of the characteristics of the cases that were found to be research fraud.

CONFIRMED CASES OF RESEARCH FRAUD: SUBSAMPLE DISTRIBUTIONS AND A PATTERN SUMMARY

Our data show that in 358 of the suspected cases clear research fraud was found. Since readers may wonder whether that analysis of only these cases can add to our understanding of fraud in research patterns, we reanalyzed the distributions of relevant variables, but we discuss only the variables on which we had sufficient information and avoid the variables where the numbers are too small. The purpose of this is to shed focused light on the emerging patterns involved in fraud in research.

The cases in which fraud was found were mostly conducted by individuals (339 or 94.7 percent of the cases). The offenders were mostly male researchers (75.1%). For the 247 cases in which we know the academic rank of the offender, the largest group (35.2%) were senior advanced professors with tenure, 19 percent were postdocs, 15.8 percent were research staff members, and only 10.1 percent were untenured academic researchers (the rest were laboratory staff and research students). Most of the cases involved researchers associated with universities (82.2%) while the rest were affiliated with university hospitals, general hospitals, and private or public research organizations. By discipline, most research fraud took place in medicine (56.7%), then in the natural sciences (32.9%), while 10.4 percent of the cases were from the social sciences or the humanities.

With regard to the funding sources, of the 310 cases of confirmed research fraud for which we have information on the funding sources, 296 cases were studies that either had public funding or were detected in the process of applying for public funding (95.4 percent of the cases). We also know that in most detected research fraud cases, the research was funded (72.6%) and in 59.5 percent of the cases that were confirmed as involving fraud, the research had already been published.

Regarding the character of research frauds, the most dominant category was of falsification of data (57.5%), then in descending order, fabrication (32.9%), plagiarism (22.8%), misrepresentation of data (14.8%), and ethical issues with regard to research (3.9%).[44]

Regarding the opening of the investigation, the largest category of verified research fraud cases began because of complaints by collaborators (28.5%). Then, in descending percentages, investigations were initiated by suspicions raised by colleagues from the same institution (11.4%), followed by the mentor (9.8%) and by reviewers of the funding agency (5.7%).

In almost all cases in which suspected research fraud was confirmed, the institution in which the suspected researchers were employed made an investigation (336 cases, 98.8%). This assuring finding indicates that employing organizations are taking suspicions seriously.

Finally, in 48 cases the organization fired the accused researcher and in 56 cases the researcher resigned. In only 135 cases in our database did the accused researcher admit involvement in fraud. This is only 38 percent of the cases in which research fraud was confirmed through investigation. This figure indicates not only how complex such cases are, but that admission of wrongdoing is not an obvious outcome of verified cases of fraud.

How do the patterns of the distributions of confirmed research fraud correspond with those of alleged or suspected fraud in our data? Although the patterns are similar, there are some instructive differences between what starts as a suspected case and what ended as a confirmed case on research fraud. While being cautious over the partial data we have on cases, we observe that more confirmed than suspected cases of fraud were by individuals (94.7% vs. 87%), by male researchers (78.3% vs. 75.1%) and by senior professors with tenure (41.5% vs. 35.2%). A higher percentage is of confirmed cases of falsification (57.5% vs. 51%), but less for fabrication (32.9% vs. 38.7%) and for plagiarism (22.8% vs. 25.6%).

The confirmation of fraud versus the earlier stage of suspicion happens in a higher percentage when the initial complaint is by collaborators (28.5% vs. 21.2%) and when it was raised by colleagues from the same organization (11.4% vs. 9.7%).

Although comparing the patterns in the subsample of confirmed cases of fraud to the whole sample of suspected cases of fraud should be taken with caution, some conclusions can be specified. Confirmed cases of fraud were in higher percentages for individuals, males, senior tenured professors, and for falsification cases. Detection by collaborators and colleagues seem to increase the probability for confirmed research fraud.

THREE | An Organizational Approach to Research Fraud

As cases of research fraud mounted and became more complex, and because of the uproar such cases evoked in the public sphere, institutions began paying more attention to the problem. Examining the organizational approach to cases under the umbrella name "research fraud" has thus become an interesting and important issue. How indeed have organizations reacted to various allegations of research fraud?

BACKGROUND

History teaches us that what has been referred to as research fraud is not a new phenomenon. The earliest case recorded in our database is from 1891. Since then (and probably before then), there have been many cases of research fraud, in which scientists plagiarized ideas or data or, alternatively, invented, copied, or manipulated data so that it would match the findings they wanted. While many scholars have studied these phenomena, their emphasis has been on the scientists or the disciplines, but not on the organizational context in which these deviant scientists practiced.

In this chapter we will apply some organizational lenses to the study of research fraud. We will look at the organizational context in which the fraud occurred, at organizational deviance as routine nonconformity, and at the organization of conducting scientific or other scholarly work and its relations to research fraud.

One of the first claims of early institutional theory was that it is important to distinguish between the terms "organization" and "institution." In the literature of organization theory, this distinction was first introduced in 1949 by Philip Selznick, an organizational sociologist from the University of California at Berkeley, who claimed that when formal organizations become "infused with value," they are institutionalized (i.e., become institutions). Institutionalization is the process by which culture and history cement the social order and constrain organizational choices, which leads to patterns that represent normative and cognitive fixity, and become taken for granted (Meyer, Boli, and Thomas 1987, 13).

In one of the founding papers of new institutional theory, DiMaggio and Powell (1983), two prominent organizational sociologists, discuss the process through which organizations become institutionalized. They claim that organizations conform to pressures from their environment to adapt to normative expectations in order to gain legitimacy, which is crucial for the continuous success and survival of the organization. DiMaggio and Powell highlighted three processes by which organizations adapt to such pressures: coercive, normative, and mimetic. The coercive processes result from political pressures and the force of the state, which imposes regulatory control. The normative processes come from the strong influence of the professions and the role of education systems, while the mimetic processes result from the organizations' habitual responses to uncertainty. When research-related organizations comply with these pressures, they gain high levels of legitimacy and as a result are treated as "institutions."

Another central and related aspect of institutional theory claims that in order to understand organizations and their actions and forms as they result from external pressures, scholars should apply the "organizational field" approach (DiMaggio and Powell 1983). "Organizational fields" are defined as "sets of organizations that, in the aggregate, constitute a recognized area of institutional life; key suppliers, resource and product consumers, regulatory agencies, and other organizations that produce similar services or products" (1983, 148–49). The organizational field generates the institutional life and applies pressures on organizations to adapt to normative expectations.

These basic tenets of institutional theory are central to our discussion of research fraud. First, it is important to note that the environment in which research-related institutions operate is constantly pressuring them to conform to the external norm of treating seriously issues related to research integrity. One of the key issues that new institutionalist scholars further try to explore is whether this conformity has to be real or can merely be ceremonial. What matters here is not what organizations do but what others in the field perceive it to be doing. This can explain the façade behind which normal business goes on uninterrupted when an organization is perceived to be seriously dealing with research integrity. Specifically, we claim that all stakeholders related to the conduct of research work (e.g., universities, public and private research funding agencies, scholarly journals, professional associations) are subject to external pressures to conform to the increasingly stringent norms for monitoring research and insuring integrity, and that they act to be perceived as serious about it. Second, some types of organizations are perceived to be institutions in the sense described above. They become "taken for granted" in society (Meyer and Rowan 1977) and this means they are perceived to be highly normative and of great value to society, and no one doubts their importance and legitimacy.

THE ORGANIZATIONAL CONTEXT OF RESEARCH FRAUD

In order to add the organizational lenses to our consideration of research fraud, we start with the question: What does the general term of "organizational misconduct" imply? The organizational theorists Greve, Palmer, and Pozner (2010, 56) define organizational misconduct as "behavior in or by an organization that a social-control agent judges to transgress a line separating right from wrong; where such a line can separate legal, ethical, and socially responsible behavior from their antitheses." This definition treats misconduct both on the organizational and the individual levels and views it as a behavior of individuals within organizations but also as a behavior of organizations themselves.

This chapter started by describing the external normative pressures on organizations in order to understand how and why organizations comply with the process of being perceived as institutions. Now the focus shifts to the organizational context in which fraud takes place and how organizations detect and treat it.

Greve et al. bring to the fore the importance of social-control agents in treating organizational misconduct. A social-control agent is "an actor that represents a collectivity and that can impose sanctions on that collectivity's behalf" (Greve, Palmer, and Pozner 2010, 56). Since this definition focuses on intraorganizational misconduct—the acts of fraud that are conducted within specific organizations—the conduct and its treatment are specific to the organization in which the misconduct occurs. Therefore, in organizational misconduct, the actions of social-control agents are idiosyncratic to the specific organization in which the misconduct takes place. Further, Greve, Palmer, and Pozner (2010) argue that the judgment by a social-control agent is crucial in the definition, because it allows for examination of when the boundary line separating right from wrong is invoked by the agent. We add to this argument that the social-control agents also act as moral entrepreneurs in their own self-interest. This may become a major obstacle to reforming research ethics regulation, which includes various personal and economic interests invested in the revised regulatory systems. The fact that we are now finding more cases of research fraud may be the result of the increased scale of social-control agents and their awareness and interests. This is similar to the well-established relationship between increasing numbers of police officers and increasing amounts of reported crime simply because there are more people looking for crime and trying to justify their jobs by recording it (Marvell and Moody 1996 found an impact of police on most crime types to be substantial).[1]

Since social-control agents are the critical factor in determining and punishing fraud, Greve, Palmer, and Pozner do not view the occurrence of such cases as a predictable linear outcome of a set of laws, ethical principles, and/or social norms. This means that by knowing the laws, norms, and ethical principles in a set of organizations, one can still NOT be able to offer linear predictions as to where, when, and what kind of misconduct cases will take place. Choosing this approach opens a new way of perceiving misconduct, as constructed by the actions of social-control agents and their impact, beyond organizational laws, norms, or ethical principles.

According to Greve, Palmer, and Pozner (2010), there are two main benefits of using this approach. First, it makes the identification of misconduct an empirical question where empirical evidence determines the extent to which a behavior is designated as an instance of wrongdoing by social-control agents. Second, this definition allows studies to analyze the importance of an observable action to agents such as managers in an organization.

A similar approach is offered by Palmer's work on wrongdoing in a normal organization (2012). In his provocative work, Palmer offers a distinction between organizational wrongdoing as an abnormal phenomenon and as a normal phenomenon and shows that different theories approach organizational wrongdoing using different assumptions that allow for misconduct to be normal at times and abnormal at other times.

According to Palmer, the distinction between right and wrongdoing is the result of two elements. One is the judgment of a social-control agent or the whistle-blower, those who monitor and deal with wrongdoing and those who detect such cases and expose them to the public. These can be managers or employees in organizations or teachers in schools. In the case of our study on research fraud these agents include scientists, department chairs, faculty deans, committees in universities or research funding agencies, and editorial boards. The second element includes the actual actions taken. Thus, the way organizational "wrongdoing" is constructed by the social-control agents has an impact on what actions are taken by the organization after "wrongdoing" is detected.

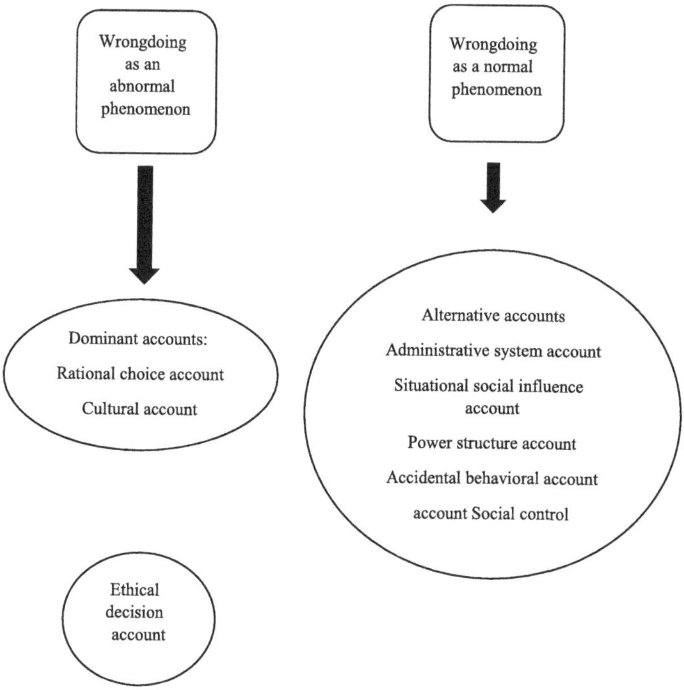

Fig. 3.1. The two perspectives to organizational wrongdoing (based on Palmer 2012)

THE "ORGANIZATIONAL FIELD" APPROACH TO RESEARCH FRAUD

As noted above, our analysis is using an "organizational field" approach to research fraud and claims that there are various organizational stakeholders that act as "social-control agents" who detect, monitor, or investigate cases of misconduct. Based on our organizational field approach to research fraud there are multiple institutional interested stakeholders that are external to universities where most research is conducted. These can act as interorganizational social-control agents that apply pressure to detect and punish research fraud.

We therefore claim that the field of research fraud is different from single organizational wrongdoing where the misconduct has mainly internal implications. The "organizational field" (DiMaggio and Powell 1983) approach to research fraud depicts a network of institutions acting as stakeholders who have a joint interest in securing research integrity and preventing research fraud. Thus the dynamics of internal methods of preventing research fraud in universities, for example, is interrelated with the field pressure of other institutional agents such as government funding agencies (NIH for example), private funding agencies, editorial boards of journals (*Nature* for example), and scientific associations. All these institutions play a key role in the negotiation process of detecting and sanctioning research fraud cases and increasing the awareness of the scientific community to the norms of scientific integrity.

Applying a system approach, in the form of an organizational field, to the study of historical changes in a scientific community was also proposed by Leahey and Montgomery (2011). They reviewed historical changes in the nature of professional autonomy of scientists and research regulations and described shifts from strong dependence on self-regulation of scientists, moving to heavy reliance on external regulation and then to a hybrid model of self-regulation and external regulation. In the current hybrid mode, they argue that the environment of scientific work is regulated through a blend of renewed emphasis on the norms of unmonitored trustworthiness coupled with external research regulation. In their view, this blend is facilitated by the central role of the Office of Research Integrity (ORI) that created and disseminated a new behavioral code for research activities titled "responsible conduct of research" (RCR), which expanded both self-regulation and institutional regulation.

Another claim offered by Leahey and Montgomery (2011) is that the regulatory system is composed of a range of organizational actors (constituting the organizational field) that have an impact on the focal actor, the academic researcher, and the focal activity, the research project. They describe several concentric circles that depict the relations of the organizational actors to the focal actor. In the center circle, there are the direct relations among those who are directly involved in performing the research (e.g., research assistants, laboratory technicians). The second circle includes the actors that are directly engaged in the evaluation of particular research projects (e.g., journal editors, funding agencies, collaborators). The third circle is composed of those with indirect relations to the focal actor and activity, who are engaged in setting standards and disseminating norms for the conduct of research (e.g., research universities, ORI, peers, professional societies), and the fourth circle includes those who are involved in general policymaking and mass communication. The contributions of Leahey and Montgomery (2011) to our understanding of the organizational setting of research fraud and scientific integrity are most valuable. First, they claim that the norms and control over scientific conduct are changing over the years as new actors enter the organizational field and change the regulations and norms concerning the conduct of ethical scientific research. Second, they depict the organizational field as changing as recent trends in higher education such as translational research (university-industry technology transfer) bring into the research field of scientific integrity new actors from industrial and clinical settings. Finally, by bringing in the discussion of academics' professional autonomy (or lack thereof), they illustrate how the notion of "control over the scientific work process" is added to the two traditional concepts of "control over work conditions" and "control over content." These new regulations and control processes require scientists to archive data, provide access to data, or describe in great detail how they conduct their research, which adds major administrative constraints to scientific work.

Similar to the argument presented in this chapter, Montgomery and Oliver (2017) suggest using the institutional theory approach to organizations (Meyer and Rowan 1977; DiMaggio and Powell 1983) as the framework for studying how and why organizations and individuals act as they do. In their terms, the "main premise in institutional theory is that environmental forces play a key role in shaping behaviors within an organization. That is, an organization's actions are not independent from, or immune to, forces outside its boundaries, but rather are constrained by the influence of norms, beliefs and rules in the environment" (Montgomery and Oliver 2009, 138). More on this issue later.

RESEARCH FRAUD AND LEVELS OF ANALYSIS

Organizations—universities, research centers, hospitals, or knowledge-intensive firms—are the hosting institutions of scientists. Therefore, acts of research fraud are usually committed by scientists who are affiliated with research-related institutions. At times, misconduct can take place between institutions, when two or more scientists that are affiliated with two or more different institutions collaborate in the research fraud.

High-quality scientific research is of value to the hosting institution with which the scientist is affiliated. In the case of universities and research centers, the output of their scientists can result in major scientific impacts and affect the reputation and ranking of the organization within the academic world. In the case of knowledge-intensive firms in which science-based intellectual property is developed for making products or providing services that are brought into the marketplace, the scientific work conducted there is a crucial part of the commercial process. In hospitals, high-quality medical research adds to the reputation of the hospital and the doctors who work there.

Yet even if a study of research fraud focuses on the institution/organization perspective, the unit of analysis does not necessarily have to be the organization. The analyses can be conducted on different levels or on an integration of multiple levels. The organizational-level analysis can focus on the independent work of scientists within the university or the firm, thus concentrating on single individuals conducting scientific work in organizations. The analysis can also focus on small organizational subunits such as research laboratories where students, research assistants, or laboratory technicians are working on studies with their advisors or the heads of the laboratory. In addition, the organizational perspective can focus on multiorganizational research arrangements, collaborative scientific work between scientists working in two or more different organizations. These can include two or more universities, two for more firms, or a university and a firm. Of course, other combinations can exist as well.

RESEARCH FRAUD AND SOCIAL-CONTROL AGENTS

Social-control agents within organizations contribute to the norms that assist in the prevention of research fraud. But the situation becomes more complex when there are a number of varied stakeholders involved in the research. This complexity results from the multiple stakeholders who are involved and the multiple levels of analysis that are relevant to cases of research fraud.

Actors acting as social-control agents can be from all levels of the institution. They can be members of a research team, thus acting from within the group, they can be colleagues on the department or faculty level, such as a research team leader, colleague in the same research area, or a manager in an organization, or they can be top university administrators such as a rector or a president. In addition, due to the involvement of agents from different institutions in the organizational field in monitoring and sanctioning or punishing research fraud, there can be collaboration between control agents who are affiliated with different organizations yet share an involvement in a specific case of research fraud, such as journal editor and a university rector.

Table 3.1 describes the different levels of analysis and four types of social control agents who are acting as central stakeholders in the field of research fraud. These include universities, knowledge-intensive firms such as biotechnology firms, academic or state-funded research centers, and academic journals. The cells in the table depict possible positions of specific social control agents.

There are several elements that appear in the organizational literature on research fraud that are relevant to our interest in it as an organizational phenomenon. First, different kind of organizations experience different levels of pressures from their institutional or technical environments (Meyer and Scott 1983). While the technical environments are those in which organizations produce products or services to the market, institutional environments are those in which rules and requirements pressure organizations to conform in order to receive legitimacy and support. The pressures of the institutional environments in which the organizations operate has an impact on the internal processes within them. Some organizations experience less pressure from their institutional and technical environments than others (Meyer and Scott 1983). For example, organizations that aim to generate technology-based products experience less institutional pressures from their environment than do hospitals or schools, but more pressures from their technical environment. In the case of universities and other research-based institutions, there are great institutional environmental pressures to comply with norms of research integrity and to punish breeches of these norms. Compliance with such norms can increase the sensitivity of scientists and university administrators to research fraud and generate pressure to be intolerant with scientists who do not comply with the norms of scientific integrity. Thus, conforming rigidly to environmen-

Table 3.1. Social-Control Agents for Research Fraud by Levels of Analyses

Organizational Level / Social Control Agent	University	Knowledge Intensive Company	Academic/ State-funded Research Center	Academic Journal
Group	Laboratory research team	Laboratory research team	Laboratory research team	Reviewers, editorial team members
Department	Department chair	Department manager	Head of research team	Editorial board
Organization	University top management committee	Firm top management	Head of research center	Chief editor
Organizational collectives	A group of universities	A group of firms	A group of research centers	Collection of journal editors

tal pressures regarding the prevention of research fraud and treating detected cases harshly can intensify the processes of detecting and sentencing research fraud within organizations.

Second, the general organizational literature has found interrelations between the degree to which managers in organizations are perceived as expressing weak ethical behavior and acting in unjust ways and the level of research fraud that is detected within the organization. A study by Currell and Tracy (2010) describes a survey conducted by the Corporate Executive Board's Compliance & Ethics Leadership Council (CELC). According to this survey, there were more serious types of fraud when employees felt that their superiors expressed a weak ethical culture and conducted unjust practices; these findings are similar to the findings of Barber et al. (1973). Thus, the employees' perception of organizational culture and the levels of fraud were positively correlated. This means that the ethical culture of the institution in which individual organizational employees practice seems to be an important factor in whether or not a scientist commits organizational fraud. The study also found that a perception of high corporate values leads to a more ethical work environment. This argument can be applied to the field of research fraud to predict that in universities with strong codes of ethics coupled with strong values and a culture of scientific integrity, there will be fewer cases of research fraud.

DIMENSIONS OF THE ORGANIZATIONAL CONTEXT

The organizational context matters for understanding and analyzing research fraud in several different ways. To begin with, organizations create codes of ethics and apply them to the work of their scientists through organizational bureaucracy instructions or socialization processes. That is, we can view organizations as enforcers of ethical codes. These codes can express internal norms or can result from external pressures of other organizations or institutions to apply restricting codes to science workers. For example, funding organizations such as the NSF or NIH in the United States impose clear requirements on universities prior to funding the research proposals of their scientists. Scientists working in universities that do not establish such strict internal requirements and codes of ethics will not have access to such funding. The development of these codes and their dissemination to all the scientists working in an organization may prevent misconduct in the organization and constitute the mechanisms for generating socialization to the norms of scientific integrity among all newcomers.

Organizational sociologists Kathleen Montgomery and Amalya Oliver (2009) discuss the issue of changes in "institutional logics" and describe how the institutional logic in the field of research fraud has changed over time. In general, institutional logics are the belief system and associated practices that predominate in an organizational field (Scott et al. 2000: 170). The institutional logics define the organizing principles and practices that are the guidelines for the field's members (all different organizational actors that have an interest and play a role in an interorganizational domain). These logics are phrased in the form of formal language expressed by formal agencies with regard to their perception of research fraud.

Public agencies have always been the main source of funding supporting faculty research in U.S. universities. Montgomery and Oliver (2009) show in their study that in the past these agencies expected that implicit norms of scientific conduct would guide proper scientific behavior and that socialization to normative practices would prevent research fraud. Information about cases of research fraud, however, changed the expectations that guided the practices of these agencies. As the publicity about research fraud increased in the late 1960s and early 1970s and peaked in the 1980s, it triggered a shift. The change was from implicit norms to explicit behavioral proscriptions, while the guidance process was facilitated by stronger bureaucratic and administrative supervision. Later on, as private sources of research funding started exerting new pressures on research behavior, new public-private partnerships emerged to articulate explicit, yet voluntary (yet pressuring for conformity) prescriptive norms of research.

The analysis conducted by Montgomery and Oliver (2009) was of documents referring to research fraud over time. The results show the strength of influential actors in the field of research that created changes in the norms of research fraud and integrity. They found that changing the norms of acceptable conduct and integrity led to clearer, more explicit sets of expectations that focused on the centrality of "scientific integrity" while conducting research. These changes have occurred through encouraging journal editors to articulate uniform requirements for the scientific papers submitted to their journals.

Thus institutions are the context in which practices of research fraud can be reacted to, avoided, and prevented. This is because these institutions are capable of creating and operating social-control mechanisms as well as giving scientists advanced notice and training on issues of integrity and how critical it is to avoid research fraud. The organizational context in terms of prevention language has shifted over the years as different logics and values regarding conduct and norms of scientific integrity have changed. In fact, examining the attitude change regarding research fraud is but a reflection of the changes that have taken place in the operation and practice of science.

Second, organizations structure the research environment in which scientists pursue their scientific work. This research environment can be formally designed, where scientific teams and research tasks are defined by the goals and needs of the organization and by the instructions of the team leaders. This may be the case in knowledge-intensive organizations or some specific research institutes.

On the other hand, within universities, the scientific tasks or expectations tend to be less predesigned and the scientists are generally given autonomy in conducting research in their areas of interest after they have raised funding and established expertise. In such cases, the institutional expectation is that the research outcomes will be of high impact and make important contributions to relevant scientific disciplines and to the reputation of the department and the university as a whole. Consequently, research-oriented institutions are important creators of the environment in which the scientific work is conducted and where research fraud can result from related institutional pressures.

Third, the organizational context matters when suspicions are raised that a scientist has violated the ethical norms of conducting research. In these cases, the hosting institution(s) have to decide how to react to such information. For example, they have to decide whether to launch a full-scale inquiry (e.g., by appointing an investigating committee), interview the laboratory head or some colleagues to get some initial information, or try to avoid the issue altogether. Such questions as what information to disclose to the public and how to diffuse the information about the event to other scientists within the institution are always there. Finally, and this is indeed a prime challenge, the host organization has to decide upon the formal procedures to be used and the range of punishments they wish to apply in case the suspicions are found to be valid.

In summary, organizations are important actors who impact the integrity of the research they sponsor and the commitment of scientists to ethical conduct of research. They provide the context that is establishes the norms of ethical research behavior and deals with suspicious cases or signals of misconduct and relevant complaints. They are also responsible for investigating cases of research fraud and if necessary bring them to trial where punishment may be delivered. Eventually, these organizations are vested with the responsibility of informing members of the organization, the general public, and other stakeholders in the organizational field of the findings of the investigation and its resulting outcomes.

The issue of stakeholders is also important in the context of research fraud, since there are multiple stakeholders that may express interest in preventing or punishing research fraud. Aside from the general public and the media, there are various stakeholders that are relevant to such cases. These include the scientific journals where the scientists accused of research fraud may have published bogus research, research funding agencies who supported such research, scientific associations to which the scientists belong, and other scientists in the community who may have relied on scientific work that was found to have been fraudulent.

ORGANIZATIONAL DEVIANCE AS ROUTINE NONCONFORMITY

In an important essay, Diane Vaughan (1999) provides an overarching theoretical and conceptual presentation for the field concerning the "dark side of organizations," which includes mistakes, misconduct, and disaster. Her analysis focuses on formal organizations as units designed to produce means-end social action through formal structures and processes and aimed at producing reliable, replicable, and goal-oriented outcomes. Following this definition, universities, for example, are formal organizations designed and structured to produce high-quality reliable and confirmatory research.

In this context, organizational deviance is defined by her as "an event, activity, or circumstance, occurring in and/or produced by a formal organization, that deviates from both formal design goals and normative standards or expectations, either in the fact of its occurrence or in its consequences, and produces a suboptimal outcome" (1999, 273).

This definition is broad, as it captures the occurrence within a formal organization and treats the deviation from the "formal design goals" and the "normative standards or expectations" as leading to a suboptimal outcome. The discussion is on the organizational level, but the actions are being performed by individual members of the organization.

Vaughan includes in this definition three main components: mistake, misconduct, and disaster, and points out that they differ from each other not in the general fit with her definition but in their particulars: "the normative standards or expectations violated (internal rules; legal mandates; social expectations), the categories of the public harmed, the extent of harmful social consequences, the social group whose expectations are violated, and/or the response of some public to the incident" (1999, 273).

Vaughan describes the main characteristics of organizational misconduct, as distinguished from mistakes or disasters, as "acts of omission or commission by individuals or groups of individuals acting in their organizational roles who violate internal rules, laws, or administrative regulations on behalf of organization goals. The extent of adverse consequences and harm to the public will vary with the act, so social cost may be contained or diffused" (1999, 288).

The environment plays a significant role in regulating organizational misconduct. Vaughan claims that the role of the regulatory environment in the social origin of misconduct is effective only in specific cases. It is based on issues of power of both the structure of the regulatory environment and specific regulatory outcomes. In following Robert Merton's claims (1968a), the hypothesis is that organizations with limited access to economic opportunities are those that will be more likely to commit acts of misconduct.[2]

Following this general focus on resources, more recent sociological qualitative research that analyzes cases of organizational misconduct by nonprofits, governments, and businesses shows that misconduct is not limited to marginal and failing organizations. Instead, competition over various kinds of scarce resources (and not just financial resources) seems to be crucial both to the survival of the organization and to the incidences of misconduct (Vaughan 1999, 288–89).

Another aspect of the environment, less emphasized earlier, is culture. Since the conceptualization of culture refers also to the normative environment of organizations, it is important to understand the overall normative elements in the organizational field that directly and indirectly affect the level of ethical conduct within organizations. Palmer (2012, 66) argues that the cultural explanation of organizational wrongdoing is a dominant one. Cultures in organizations can facilitate wrongdoing in two ways: by endorsing wrongdoing and by arguing extenuating circumstances. The first of these mechanisms can give rise to wrongdoing by creating the norms, values, beliefs, and assumptions that treat specific or general wrongdoing as rightful actions. The second mechanism is based on organizational assumptions, values, and beliefs that make wrongful behavior seem understandable and acceptable. Palmer (2012, 72–80) suggests six types of neutralization that are used as after-the-fact excuses for wrongdoing and that protect the wrongdoers from the consequences of their actions. These include denial of responsibility, denial of injury, denial that there is a victim, social weighting (making the assumption that the wrongdoing is no more or less severe than others' wrongful actions), appeal to higher loyalties, and balancing the ledger (accepting wrongdoing if the person acting wrongfully is doing or plans to do good deeds that compensate for the wrongful act). This argument follows the initial theoretical framing of "account" as explanations of wrongdoing by Scott and Lyman (1968).

It is also instructive to examine how the competitive, regulatory, and cultural environments affect organizational processes and lead organizations to create a normative environment of scientific integrity that prevents misconduct or treats it harshly when it is detected.

Vaughan (1999) argues against a unidimensional view in which culture is treated as a unified given within organizations, and she suggests a theory that emphasizes cultural complexity and variation and even offers a view of conflicting cultures when different members or professions within organizations hold different views of the organizational culture.

Vaughan suggests that scholars of organizational misconduct should integrate into their studies three levels of analysis: aspects of the environment in which the organization operates, characteristics of the organization (structure, processes, and tasks), and cognitive practices and perception of misconduct. Such analyses, which are usually associated with the bright aspects of organization, are by definition also related to the dark side and should be studied in the same manner.

The distinction between the bright side and the dark side leads to an interesting research question: When do organizations produce proper conduct and when do they produce misconduct? In our study, this question is of interest with regard to research fraud, yet it is hard to specify with regard to proper conduct, since research fraud cases are detected sometimes, yet proper conduct may always include cases of research fraud.

RESEARCH FRAUD AND THE ORGANIZATION OF SCIENCE

The third aspect related to how organizational elements are associated with research fraud involves the organization of science, and by this we mean the way scientific work is conducted and organized in different disciplines. For example, is the scientific work conducted by individual scientists or teams of scientists, by disciplinary or interdisciplinary research teams, by only scientists with PhDs, or by a mixture of undergraduate students and graduate students? If the research is conducted by a team, questions arise as to the structure of the team's research work (joint or independent), the role that the team leader takes, and the degree to which team members share their findings and discuss their research problems among themselves and with others.

Moreover, there are additional issues that involve the organization of the scientific work in different disciplines. One important question is related to the degree to which it is common to replicate scientific work. In some disciplines, research replication can be characterized as a difficult-to-impossible task. For example, in the discipline of social anthropology, it is hard to replicate studies because the interaction between the anthropologist and the observed field becomes a part of the study's findings. On the other hand, in life sciences or the physical sciences studies, it is much easier to replicate as long as all components are carefully controlled for, since the scientist's involvement in the study should not change the results of the experiments.

Another issue related to the replication in different disciplines is whether the discipline rewards scientific work that aims mainly to verify and validate findings through replications. In general, organizing research around replication studies is of great value for the progress of science, because it has the authority to verify the validity of scientific findings. But not all disciplines reward replication-based scientific work, and not all journals publish such studies.

Clearly, one of the most important organizational methods of avoiding, preventing and detecting research fraud is through replication-based research. Many times, only consistent efforts to replicate research where novel findings are described allow for the detection of research fraud, but even when replication fails, it is hard to prove that intentional research fraud took place.

A recent example shows how the failure to demonstrate replication is debated. On July 10, 2013, there was a column in *The Scientist* magazine describing a case in which the inability to replicate reported scientific results were a basis for suspecting research fraud:[3]

> Dmitry Kuznetsov, a Russian biochemist who studies magnetic isotope effects on enzymes, has left his posts as the chief editor of the *British Journal of Medicine and Medical Research* and the *International Research Journal of Pure and Applied Chemistry*. The change was prompted after an article by *The Scientist* exposed allegations that Kuznetsov had committed numerous instances of scientific misconduct.

The allegations included inventing citations and materials. Kuznetsov maintained that he did nothing wrong. In a lengthy e-mail to *The Scientist*, Kuznetsov explained why he thinks others were unable to replicate his results. "Actually, nothing is strange about technical difficulties in enzyme function studies which is a pretty complex 'cooking'—it is not always easy to reproduce (reconfirm) somebody's data." As for researchers' inability to identify some of the citations Kuznetsov references in his papers, "my only—but no doubt terrible!!!—mistake was that I was naïve enough to publish in the West-appearing periodical the references to articles originally published in Soviet / Russian as well as in Serbian, Bulgarian . . . journals in native languages (Russian, Ukrainian, Serbian, Bulgarian . . .)," many of which ceased publication in the 1990s. Kuznetsov said he is "sick and tired" of explaining why others cannot verify his work, and "as a matter of fact, I gave up answering such criticisms about 10 years ago."

The issue of replication of findings as means of corroborating research results is a complex task that has many objective obstacles. It has been of great interest in psychology research since it is hard to compare samples, similarly control the environment, and avoid cultural differences. Beyond these complexities, replication-based research is not common among researchers since it is hard to publish such research. As described in the *Guardian*[4] on September 13, 2012, "This issue is exacerbated in psychological research by the 'file-drawer' problem, a situation which characterizes when researchers who try to replicate and confirm previous studies find it difficult to get their research published. Scientific journals want to highlight novel, often surprising, findings. Negative results are unattractive to journal editors and lie in the bottom of researchers' filing cabinets, destined never to see the light of day."

Thus, although replicating research findings is an important tool for detecting research fraud, there is little motivation for scientists to do such research because of the organization of science that encourages original research in which the main hypotheses are confirmed.[5]

ORGANIZATIONAL WRONGDOING

In a recent integrative theoretical framing of "organizational wrongdoing" discussed earlier in this chapter,[6] Palmer (2012), an expert on organizational theory and wrongdoing from the University of California–Davis, offers two provocative conclusions on what he terms "wrongdoing." First he states, "I came to the conclusion that even the most ethical, socially responsible, and law-abiding people are at significant risk of becoming entangled in wrongdoing when placed in an organizational context" (23). This claim indicates a provocative "normalized" view of wrongdoing in the sense that it does not see this as necessarily a deviant behavior.

Another subject Palmer explores is the issue of applying a "power balance" framework to organizational wrongdoing. In Palmer's words, "[t]he position of the line separating right from wrong is a political product, a reflection of the balance of power among competing interests rather than a representation of moral imperative. . . . That line is eminently disputable and perhaps inherently suspect" (264). In other words, in his view the definition and treatment of misconduct or wrongdoing in organizations is a result of the "balance of power between competing interests." In this context, defining misconduct is one way to sustain the existing structures of recognition and reward. There is no clear and unified definition of organizational wrongdoing that is independent of the power balance relations within the organization.

If we accept Palmer's argument that any organizational member is "at risk" of committing misconduct, the questions that follows is: What are the conditions that may lead organizational members to commit misconduct?

One answer can be offered by Vaughan's work on misconduct (1998). She argues that the competition for scarce resources encourages institutions to commit various forms of misconduct. Universities are willing to violate scholastic regulations in order to recruit field winning athletic teams; police departments are willing to violate the law in order to make arrests that bring recognition and funding; and political parties and governments are tempted to commit illegalities in order to gain national and international power. Similarly, she claims, in order to win grants and general prestige, scientists within universities are tempted to falsify results.

In a review of Palmer's book, Karl Weick (2013) a prominent organizational sociologist, argues that wrongdoing can be perceived as "normal" as we move away from the dominant organizational perspectives of "rational choice" theory and the perspective of "organizational culture." These two approaches treat wrongdoing as abnormal. Then, as we move to the alternative perspectives such as of "administrative systems" and "situational social influence," to the "power structure" or "accidental wrongdoing," it seems apparent that bounded rationality[7] and social influences, coupled with competition for resources, may lead organizational actors into acting wrongly. Yet these cases will not be defined as abnormal.

STAKEHOLDERS' APPROACHES TO RESEARCH FRAUD

As research fraud in science has become a topic that is widely discussed in public, different stakeholders have taken a strong and nonnegotiable stand on the issue. As mentioned above, the stakeholders range from journals, universities, research funding agencies, research centers, and hospitals on the organizational level and editors, reviewers, scientists, scientific committee members, research team leaders, and research physicians on the individual levels.

In recent years, the reference to research fraud became an important and central topic broadly discussed in articles appearing in various scientific journals.[8] This high alert level undoubtedly reflects both the genuine concern and alarm of scientists and editors over the detection of so much research fraud as well as a search for mechanisms that could avoid such cases. Due to the multiple stakeholders involved in the issue, discussions about changing policies with regard to the practice of scientific work can take place on different levels, even on the national level.

CHANGES ON THE NATIONAL LEVEL

When a case of research fraud gets a high level of attention, it may raise the call for re-evaluation of research fraud policies, since without them all scientists from that nation could be penalized by various scientific stakeholders from other countries. In this context, the journal *Cell* published a column by Colin Macilwain, editor and writer on science policy, which stated,

> In the past year . . . there has been an upsurge in global activity to address research misconduct. Scientists and editors are pushing universities and research agencies to take more responsibility for investigating and punishing misconduct. Canada has put new procedures in place to do this, and other nations are under pressure to follow suit. . . . Two main factors are likely driving this upsurge in official activity. One is the drip-drip-drip of high-profile research fraud cases. In 2005, Seoul National University unmasked the fraud by the Korean stem cell biologist, Hwang Woo-Suk, who had reported the successful production of human embryonic stem cells by cloning. Two years later, the Intergovernmental Panel on Climate Change erroneously reported false claims that all Himalayan glaciers would

likely melt by 2035. Then, in 2010, the contentious research of Andrew Wakefield—which advocated a link between autism and the measles, mumps, and rubella vaccine—was discredited when the UK General Medical Council convicted him of research fraud and questioned the content of his papers.

A second factor is the flood of hard information regarding the frequency of research fraud. The caseload of the ORI has approximately doubled over the last decade, during which time the total number of retractions in scientific literature has risen tenfold, reaching 400 last year. The emerging sub discipline of research into research integrity (RRI) has built up a body of evidence that "serious misconduct," generally classified as "falsification, fabrication, and plagiarism (FFP)" is more frequent than these numbers suggest.[9]

The article goes on to cite Daniele Fanelli (2009) of the University of Edinburgh, an evolutionary biologist who shifted to the study of research misconduct. He said that " 2% of researchers confess to FFP,[10] and as many as one-third confess to lesser forms of research fraud. Fanelli also says the work has been surprisingly well received by other scientists. "One of the things that has made misconduct less of a taboo is the fact that researchers are studying it, the same way that they would study anything else." In a later publication, titled "Why Growing Retractions Are (Mostly) a Good Sign," Fanelli (2013) follows the argument that misconduct is less of a taboo. He claims that his analysis of retracted papers shows that the increasing number of retractions is probably caused by a growing propensity to retract flawed and fraudulent papers, but that there is little evidence of an increase in the prevalence of misconduct.

Macilwain (2012, 1419) also quotes Mark Frankel, the director of scientific responsibility, human rights, and law at the American Association for the Advancement of Science (AAAS): "The community's approach has really changed. . . . They were in denial for many years: but now, with twenty or thirty people being found guilty of misconduct each year, researchers realize that there is a problem."

Macilwain further describes the new Tri-Agency Framework (2012, 1417):

> One example for the Academy to consider is Canada's new Tri-Agency Framework on the Responsible Conduct of Research. The agency is backed by the three main government-funding agencies, and it has a staff of eight to cover a research system one-tenth the

> size of that of the United States. The responsibility of the agency is to oversee all research conducted at institutions that receive public funds. "We use a much broader definition than FFP, and cover the breach of any ethics policy, and any mishandling of public funds," says Susan Zimmerman, the lawyer who directs the new secretariat in Ottawa.

Macilwain indicates that in December 2012 Canada implemented this initiative, which established comprehensive new arrangements to confront the rise in misconduct cases in science. Under the new rules, they claim "institutions have to have their own research conduct policy to cover all of their research, regardless of the source of funding." Canadian researchers who get public grants have to sign a consent form that will waive their privacy rights in the event of a "serious breach" of the framework. This initiative also gives clear authority to publish names of anyone in breach of the rules, and twenty new investigations were expected in the first year.

The article[11] continues to praise this Canadian initiative: "Canada's shiny new system sounds rather like what the rest of the world would like but isn't going to get. At a stormy meeting in London this past January, for example, journal editors and senior scientists clashed angrily with university and British government officials, accusing the latter group of complacency in their handling of research misconduct."

International matters related to research fraud are having a cross-national effect on other countries, which leads to cases where if one country fails to investigate research fraud, stakeholders in another country are committed to investigating it. The article further mentions,[12] "This meeting was prompted by the failure of the University College London to investigate the Wakefield case.[13] The meeting 'got senior people to confront the issue, which some of them had not been doing,' says Fiona Godlee, editor-in-chief of the *British Medical Journal*, which co-organized it with the Committee on Publication Ethics (COPE)."

It is clear that the detection and punishment of research fraud is perceived by the public as the responsibility of universities and of research funding agencies. If these bodies fail to investigate research fraud, however, other agencies, such as, in this case, National Institutes of Health, are getting involved with a rage. The article further indicates that "[t]he meeting was galvanized by a strong statement from Michael Rawlins, chairman of the National Institute of Health and Clinical Excellence (NICE), which is the body that determines which drugs are paid for by Britain's National Health Service. He accused the universities and their funding bodies of complacency."

Whether universities are ready to deal openly with cases of research fraud detected among their scientists is not clear. Research fraud cases can harm a university's reputation. Rawlins argues that universities are still downplaying the research fraud issue.

"There is a problem, both with fraudulent data and with the failure to publish negative results," he claims. "In the world where I live, the results of clinical trials are very important. If there is fraud going on, people will be harmed. It is ultimately down to the universities. But either they don't think there is a problem, or they think that to touch it will damage their reputations. So they want to sweep it under the carpet."[14]

SCIENTIFIC ASSOCIATIONS

Scientific associations are also responding to the increase in cases of research fraud by revisiting their policies and procedures for research fraud and by alerting their scientist members to the new explicit policies that are aimed at preventing future research fraud. For example Kohler (2012), the managing director of scholarly journals for the American Diabetes Association, published the new "Updates to Policies and Procedures Related to Potential Scientific and Academic Research fraud in the Journals of the American Diabetes Association." In this publication, she states up front, "Over the last five years, both editorial teams have not only taken great efforts to cement the status of *Diabetes and Diabetes Care* as the premier journals devoted to diabetes research, but they have also diligently worked to ensure that research published in ADA journals adheres to international standards and requirements for biomedical publications" (2012, 189).

Another example is the case of Diederik Stapel, the social psychology professor from Tilburg University in the Netherlands, who shook psychological associations around the world. The American Psychological Association publicized the case on their website, and social psychologist Jennifer Crocker, who served as chair of the American Psychological Association's Publications and Communications Board, wrote in *Nature* that "to understand fraud, we should think about how it begins and escalates, not how it ends. By the time such fraud is exposed, bad choices that would usually lead to only minor transgressions have escalated into outright career-killing behavior."[15]

The association reported on the findings of the interim report and pointed to two factors that contributed to how and why the misconduct was unnoticed for a long time. The first key element was that Stapel used a sophisticated way to exert his power and prestige. He collected so-called data in "secondary schools" and enlisted fictitious paid research assistants for data collection. The complete, coded data set was then made available to the collaborator for further analysis and manuscript writing. The second key element found by the committee was directed toward the academic community that showed poor functioning of the procedure for scientific scrutiny and criticism. The report indicated that a number of "red flags" should have been raised. These included that doctoral students did not collect their own data, that there were no completed questionnaires, that the method sections in the manuscripts were not detailed, that the data were too good to be true, and that it was not made publicly available as the culture in social psychology is.

SCIENTIFIC JOURNALS: RETRACTIONS OF SCIENTIFIC PAPERS AFTER RESEARCH FRAUD IS DETECTED

Scientific journals that publish results of research in different areas have had to act as well when published articles were found to be associated with some form of research fraud. In many cases, the articles had to be retracted and readers had to be updated in a clear and unambiguous way about the problem. An example of newspaper coverage of this topic by columnist Carl Zimmer appears in the April 16, 2012, *New York Times*. The article, "A Sharp Rise in Retractions Prompts Calls for Reform," describes the following case:

> In the fall of 2010, Dr. Ferric C. Fang made an unsettling discovery. Dr. Fang, who is editor in chief of the journal *Infection and Immunity*, found that one of his authors had doctored several papers. It was a new experience for him. "Prior to that time," he said in an interview, "*Infection and Immunity* had only retracted nine articles over a 40-year period." The journal wound up retracting six of the papers from the author, Naoki Mori of the University of the Ryukyus in Japan. And it soon became clear that *Infection and Immunity* was hardly the only victim of Dr. Mori's misconduct. Since then, other scientific journals have retracted two dozen of his papers, according to the watchdog blog *Retraction Watch*.[16] "Nobody had noticed the whole thing was rotten," said Dr. Fang, who is a professor at the University of Washington School of Medicine.
>
> Dr. Fang teamed up with a fellow editor. . . . They reached to-

> gether a worrying conclusion: that not only retractions were rising at an alarming rate, but also that retractions were just a manifestation of a much more profound problem—of a dysfunctional "scientific climate." Dr. Casadevall, who was the editor in chief of the journal *mBio*, was cited to say that he feared that science had turned into a "winner-take-all" game where the incentives lead scientists to "cut corners" and, at times, commit acts of misconduct.
>
> The two editors issued a plea for some fundamental reforms in handling research integrity. They also presented their concerns to the National Academies of Sciences committee on science, technology and the law.

The relations between the scientific community and research fraud, whether it is clear and purposeful acts or innocent mistakes in conducting scientific investigation, is further discussed in the article: "No one claims that science was ever free of misconduct or bad research. Indeed, the scientific method itself is intended to overcome mistakes and misdeeds. When scientists make a new discovery, others review the research skeptically before it is published. And once it is, the scientific community can try to replicate the results to see if they hold up."[17]

But critics like Dr. Fang and Dr. Casadevall argue that the changes in science during the recent decades are worrying, especially biomedical research, which consumes a larger and larger share of government science spending.

For example, in October 2011 the journal *Nature* reported that published retractions had increased tenfold over the past decade, while the number of published papers had increased by just 44 percent.[18] They claimed that recent retractions were a mix of research fraud and honest scientific mistakes.

That Internet-based journals now offer researchers more exposure to the scientific community is also an important factor in the severity of the problem of research fraud, both in detecting it and conducting research fraud. The article claims, "Several factors are at play here, scientists say. One may be that because journals are now online, bad papers are simply reaching a wider audience, making it more likely that errors will be spotted. 'You can sit at your laptop and pull a lot of different papers together.'"

The reputation of journals is also an important element in the retraction of scientific papers that are found to be based on research fraud. Fang and Casadevall looked at the rate of retractions in seventeen journals from 2001 to 2010 and compared it with the "impact factor" of journals, which is a score of journal prestige based on how often their papers are cited by other scientists. The two editors found out that the higher a journal's impact factor is, the higher its retraction rate. A few explanations may account for this. The most obvious is that prestigious journals are more willing to retract papers suspected of research fraud more than less prestigious journals are. An additional explanation may be that papers in such journals are subject to a more intense inspection. A much less likely interpretation could be that the quality of papers published in such journals is lower. The highest rate of retraction in their study was at one of the world's leading medical journals, *The New England Journal of Medicine.*

Dr. Fang proposed that this might be because the struggle to publish in high-impact journals may be causing increasing numbers of errors, since new PhDs, who must compete for a small number of jobs, are competing more fiercely than ever. He indicated that in 1973, more than half of biologists had a tenure-track job within six years of getting a PhD. By 2006, the figure was down to 15 percent. Recently, Smaglik (2016) reported studies claiming that only 20 percent of postdocs are likely to find secured positions.

In addition, university laboratories need a steady stream of grants from the government and other sources. Yet two recent trends, Fang claims, further intensify the competition. The NIH accepts a much lower percentage of grant applications than in earlier decades, while many universities expect scientists to draw an increasing part of their salaries from grants. These combined pressures have influenced how scientists are promoted and may further lead to lack of time to check research and to unintentional and at times intentional research fraud: "What people do is they count papers, and they look at the prestige of the journal in which the research is published, and they see how many grant dollars scientists have, and if they don't have funding, they don't get promoted," Dr. Fang said. "It's not about the quality of the research."

The final statement by Dr. Fang reflects his view that most cases of research fraud are unintentional and driven by the constraints imposed on scientists: "You can't afford to fail, to have your hypothesis disproven. It's a small minority of scientists who engage in frank research fraud. It's a much more insidious thing that you feel compelled to put the best face on everything."

To ease this harmful competition, the two editors plan on changing the rules for scientific prizes and would like to have universities take collaboration in research into account when they decide on promotions.

SCIENTIFIC JOURNALS: POLICIES ON PROMOTING SCIENTIFIC INTEGRITY

The top scientific journals are publicizing clear policies and instructions to contributors in order to promote scientific integrity in the submission of papers for review. In this section, we provide two examples from social sciences journals. The first is the *Journal of Applied Psychology*, which places an emphasis on the guidelines of the American Psychology Association (APA) Code of Ethics. This code, reproduced below, is published on the journal's web page.[19]

> Ethical Principles
>
> It is a violation of APA Ethical Principles to publish "as original data, data that have been previously published" (Standard 8.13).
>
> In addition, APA Ethical Principles specify that "after research results are published, psychologists do not withhold the data on which their conclusions are based from other competent professionals who seek to verify the substantive claims through reanalysis and who intend to use such data only for that purpose, provided that the confidentiality of the participants can be protected and unless legal rights concerning proprietary data preclude their release" (Standard 8.14).
>
> APA expects authors to adhere to these standards. Specifically, APA expects authors to have their data available throughout the editorial review process and for at least 5 years after the date of publication.
>
> Authors are required to state in writing that they have complied with APA ethical standards in the treatment of their sample, human or animal, or to describe the details of treatment.

The second example is taken from the journal *Organizational Behavior and Human Decision Processes*.[20] The journal placed on its website a message to potential contributors that requests that they review the following standards before submitting a paper for review. The title "Before You Begin" alerts potential contributors to the specific and detailed ethics requirements of the journal:

Ethics in Publishing
Duties of Authors
Reporting Standards

Authors of reports of original research should present an accurate account of the work performed as well as an objective discussion of its significance. Underlying data should be represented accurately in the paper. A paper should contain sufficient detail and references to permit others to replicate the work. Fraudulent or knowingly inaccurate statements constitute unethical behavior and are unacceptable. . . .

Data Access and Retention

Authors may be asked to provide the raw data in connection with a paper for editorial review, and should be prepared to provide public access to such data . . . and should in any event be prepared to retain such data for a reasonable time after publication. . . .

Originality and Plagiarism

The authors should ensure that they have written entirely original works, and if the authors have used the work and/or words of others, that this has been appropriately cited or quoted.

Plagiarism takes many forms, from "passing off" another's paper as the author's own paper, to copying or paraphrasing substantial parts of another's paper (without attribution), to claiming results from research conducted by others. Plagiarism in all its forms constitutes unethical behavior and is unacceptable.

Multiple, Redundant or Concurrent Publication

An author should not in general publish manuscripts describing essentially the same research in more than one journal or primary publication. Submitting the same manuscript to more than one journal concurrently constitutes unethical publishing behavior and is unacceptable. . . .[21]

> Acknowledgement of Sources
>
> Proper acknowledgement of the work of others must always be given. Authors should cite publications that have been influential in determining the nature of the reported work. Information obtained privately, as in conversation, correspondence, or discussion with third parties, must not be used or reported without explicit, written permission from the source. . . .
>
> Declaration of Interest
>
> All authors are requested to disclose any actual or potential conflict of interest including any financial, personal or other relationships with other people or organizations within three years of beginning the submitted work that could inappropriately influence, or be perceived to influence, their work.[22]

There are other cases where journals express their responsibility to be clear about their role as leading stakeholders in discussing research fraud, promoting scientific integrity, and exposing cases of research fraud and how they are investigated and finalized to their scientific community. In this context, the journal *Annals of Internal Medicine* published a paper that dealt with the issue of retracting papers from journals once they are found to be fraudulent. In this journal, the editor and another scientist published the following abstract (Sox and Rennie 2006, 611). Before presenting their statement, we want to emphasize how they call for all related stakeholders: coauthors, universities, journals, governmental institutions, and others to take an active role dealing with research fraud:

> The scientific literature is a record of the search for truth. Publication of faked data diverts this search. The **scientific community** has a duty to warn people to ignore an article containing faked data and must try to prevent inadvertent citation of it. The scientific community accomplishes these tasks by publishing a retraction and linking it to the fraudulent article's citation in electronic indexes of the medical literature, such as PubMed. This mechanism is far from perfect, as shown by a case history of scientific fraud perpetrated by Eric Poehlman, PhD. His institution notified 3 journals that they had published tainted articles. Two journals failed to retract. The third

journal retracted immediately, but other authors continued to cite the retracted article.

Another duty of the scientific community is to verify the integrity of other articles published by the author of a fraudulent article. This task falls to the **author's institution and requires** coauthors to vouch for their article's integrity by convincing institutional investigators that the suspect author could not have altered the raw scientific data from their study. Two **universities** are currently investigating Poehlman's published research. Maintaining the integrity of the scientific literature requires **governmental institutions** that have the authority to investigate and punish guilty scientists and requires that research institutions investigate alleged fraud. It **requires journal editors** to issue a retraction when they learn that their journal has published a tainted article. It requires research institutions to accept their responsibility to investigate every article published by a scientist who has published even 1 fraudulent article. Finally, it requires **authors** to take pains to avoid citing retracted articles and to issue a correction when they inadvertently cite a retracted article.

The article reviews the problem, rules and regulations, and issues of retracting fraudulent articles as well as preventing citations of such articles. This is not a simple task and the article provides the following steps that need to be taken (Sox and Rennie 2006, 612):

> What can journals do to avoid publishing a citation of a retracted article? In principle, the solution is straightforward: Journals should require authors to attest that they have checked their manuscript's reference list against the National Library of Medicine master list of retracted articles. The Library would make the task much easier if it created a Web-based program that would check a manuscript's reference list against the master list of retracted articles and send a report to the manuscript author. Journals could require corresponding authors to attest that they had checked their submitted manuscripts' references against the master list. At the 2006 meeting of the International Committee of Medical Journal Editors, the member journals changed the Uniform Requirements for Manuscripts to specify authors' responsibility to check their manuscripts for references to retracted articles, citing PubMed as the authoritative source of information for retractions.

> Journal editors have other opportunities to prevent citations of retracted articles. They could link retraction notices to retracted articles on their journal Web sites, and they could facilitate access to the full text of retracted articles and notices of retraction. . . .
>
> Authors who cite an article that proves to be fraudulent should correct the record, lest readers unwittingly propagate its falsehoods. One hundred eighty-six of Poehlman's articles are in the Science Citation Index database (Thomson Science, Philadelphia, Pennsylvania); collectively, they have 3007 citations. Someone could use the Science Citation Index database to identify each author who cited fraudulent articles by Poehlman. Contacting the authors of 3007 articles is a formidable undertaking. Who will take responsibility?

The article ends with the following statement aimed at avoiding ethical damages in the future:

> This discussion has been about repairing damage. However, *preventing* damage would save careers from ruin and save the time spent investigating allegations and assessing the integrity of articles that become suspect by association with a fraudulent author. Everyone has a responsibility to promote a culture in which research misconduct does not happen. (613)

In conclusion, this chapter aims to bring the organizations into the field of research fraud, its prevention, detection, and treatment. We started with providing a general view of recent literature on organizational misconduct or wrongdoing that suggests that misconduct can be a normal or abnormal phenomenon, depending on the theoretical lenses used. Then, by introducing the difference between organizations and institutions, we claim that it is of great importance to treat seriously research fraud as a source of legitimacy for academic and research institutions. We also suggest that the analysis of research fraud can be placed on different levels of analyses in various organizational contexts. The terms of "organizational field" and of stakeholders' roles are further developed to offer a system approach and stakeholders involvement to treat research fraud. The chapter ends with some illustrative examples of organizational stakeholders' activities and opinions on how to avoid, detect, and treat research fraud. It shows how much passion and care are expressed by the different stakeholders—including journal editors, scientists in different disciplines, and funding agencies—with regard to research fraud.

FOUR | Fraud, Lies, Deceptions, Fabrications, and Falsifications

Following the examination of cases of fraud in research and after refining the aggregate patterns of such cases, it is time to probe into the conceptual envelope within which research fraud takes place. This envelope includes a few relevant and significant issues. The logical development of the arguments in this chapter revolves around issues of lies, deceptions, and truth. First, we delve into issues of terminologies used to describe the phenomenon, sharpening our argument that researchers and research-related organizations prefer to deflect their rhetoric from "fraud" to softer terms such as "misconduct." Next we look at how many previous works on fraud/misconduct have been published, showing a clear and consistent increase in the prevalence of such works. Having clarified issues of terms used to name the phenomena and prevalence of works about it, we introduce a clarification of concepts and discuss science, research, research puzzles, and we move on to issues of change in science. After all, fraud in research does try to introduce some change. This issue forces to the surface three interesting, relevant, and important topics. The first is a distinction between unconventionalities and deviance, the second is the nature of deviant science or knowledge claims, and the third is deviant scientists, that is, our fraudsters. These distinctions are important because they imply that opening an independent space that allows legitimate and fruitful creativity is also opening the possibility of fraud. Consequently, the following logical section delves in depth into issues of deviant knowledge claims and truth, followed by two complementary sections that first discuss the nature of deceit in research (with some illustrations) and, second, discuss fraud in other comparative social contexts. These last three sections lead us into a discussion in three successive sections about deceit in cultures more generally and gives us a larger sociological comparative perspective. Naturally, these discussions lead us next to examine the nature of lies and then to focus on deviant researchers in the context of deceit, fraud, and truth. Strange as it may sound, there were researchers who committed acts of deceit and fraud out of what may appear to some to be genuine beliefs. We use a few illustrative cases to describe and discuss this curious phenomenon. The chapter ends with some general notes and reflections about the cases, discussions, and arguments that riddle this chapter.[1]

RESEARCH INTEGRITY AND MISCONDUCT

Words are not neutral, and choosing words to characterize a behavior frequently carries moral weight. In simple, unwashed words, the phenomena described in the previous chapters are focused on research scientists who lie and deceive and thus commit fraud. Nevertheless, the words that were used to describe these phenomena have been "misconduct," "wrongdoing," "intentional bias," "mismanagement," or "improper behavior." These are terms that clearly minimize and soften the implied stigma and negativity associated with terms such as "deviance" or "fraud" or even "criminal."

There are no criminal laws against lying, and as we shall see later we all lie, some more than others. Problems begin to emerge when lies hurt, damage, or are used to gain some tangible advantage. Many cultural products are based on what looks like lies proper, such as fiction books or nondocumentary movies. The distinguishing character of such cultural lies, however, is that they are known to be lies and accepted as such.[2]

Acts of fraud that are committed to gain some tangible product or advantage, other than artistic pleasure, can be processed by police criminal investigations and, if enough persuading evidence is collected, brought to trial. No one characterizes the activities of swindlers, crooks, embezzlers, cheaters, or con artists as "misconduct" (or as "intentional bias"). Such terms seem to be kept for science[3] and, if one cares to search the Internet, for police misconduct as well. Guessing why is not too demanding. These are two powerful organizations, concerned about their public image, and using the term "misconduct" to describe infractions tends to downplay the severity of the behavior involved. Interestingly enough, the organization in the United States that deals with cases of misconduct in science calls itself the "Office of Research Integrity."[4] The word misconduct disappeared altogether.

Another common denominator of these organizations—police, financial institutions, and universities—is that their operation is based on trust. Cases of fraud threaten not only their inner workings but may erode the trust that the public may feel for them. While the operation of some organizations is based on assumption of distrust (e.g., insurance), the operation of universities is based on trust as a major issue, on the personal, professional, and the organizational levels. Indeed, in 1991 Hardwig pointed out that the whole fabric and essence of research is based on and reflects trust. Moreover, because of the publicly open nature of universities, their searches for variance in research and methods, and their support of innovative ideas, they tend to attract and hire idiosyncratic persons as faculty members. To have a large number of such persons working together involves a strong belief in trust. The search for truth, a major hallmark of research in academic life, requires that those engaged in such a search be honest. Having dishonest and conniving research faculty members can cast a dark and unpleasant shadow on universities or research institutes where such members commit their follies.

One cost of using the term "misconduct" is vagueness; cases of scientific misconduct tend to present a curious collection. As noted earlier, these could include various forms of data fraud and plagiarism, ethical violations, incompetence, abuse of students and assistants, sexual harassment, and perhaps a few more. In a strange fashion, the characterization is so open that one can argue that misconduct may be such if stakeholders—using different approaches and interpretations—believe it is and act accordingly. For example, a case some may view as blatant fraud could be viewed by others as merely sloppy work. Imputation of intentions may sometimes become a thorny issue. Working within this given situation requires both flexibility and focusing. It also requires some remedy, to which we will attend later.

It is well worth noting that behavior referred to as misconduct appears in the natural, human, social, legal, and even applied or commercial fields of knowledge and disciplines. Because there are significant differences among these fields of knowledge, a better name might have been "misconduct in research" or "misconduct in academia." Nevertheless the expression "misconduct in science" or sometimes "fraud in science" have become almost standard in academic-related contexts. It is perhaps time to switch to a more accurate term such as "research fraud."

PREVALENCE OF PREVIOUS WORKS

Academic interest in studying research fraud seems to have increased exponentially in recent years. This interest reflects many issues, including the increase in the number of cases detected, a heightened awareness about the importance of preventing and sanctioning misconduct (in which the media played a role), and the wide range of organizations associated with conducting scientific work that are central stakeholders in the scientific field. Figure 4.1 demonstrates the increase in the number of articles that focus on

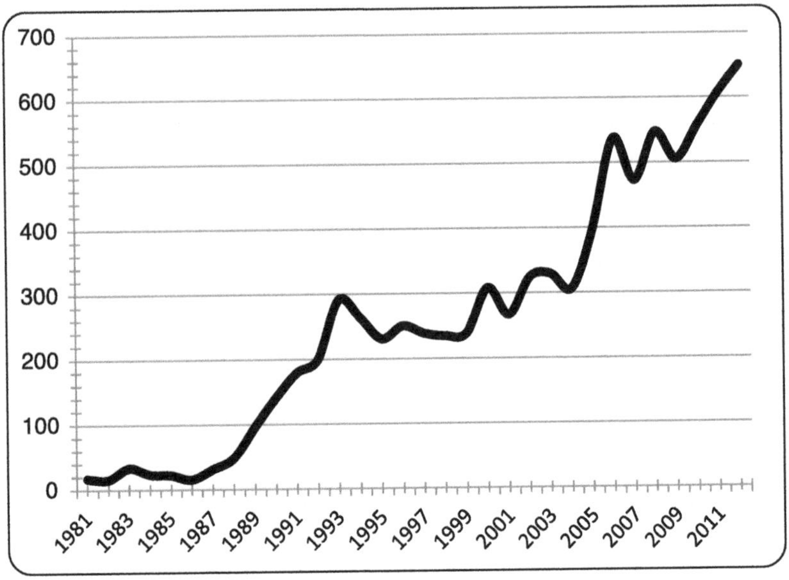

Fig. 4.1. Publications on "scientific misconduct" by numbers and years

scientific misconduct, by year, based on Google Scholar.

Due to the massive amount of scientific articles and commentary published on this topic, it is almost impossible to conduct a full, integrative literature review. In this case, reviewing articles that are themselves a review of the literature can be more beneficial.[5] Moreover, in a recent review of 115 articles on "research integrity," Macfarlane, Zhang, and Pun (2012) argue that the definition of "academic integrity" is problematic because it is open to different interpretations, including the conduct of students, mainly in relation to plagiarism and cheating. Hence, in this chapter, we aim to clarify the main relevant concepts and ideas associated with scientific misconduct and integrate them with some review of the relevant literature as well as illustrate them with some cases from our database.

CLARIFYING CONCEPTS

Science

Attempting to define or characterize science, even briefly, is truly presumptuous, and we do not pretend to exhaust the subject, either substantively or historically. For our purposes, it is enough merely to outline the main points relevant to the context of this book, as well as how we conceptualize science.

Science as we know it appeared and developed in Europe[6] from about the fifteenth century and became an accepted profession there in the late nineteenth century. There are many sciences: pure and abstract, experimental, applied, and industrial, and research scientists work sometimes alone and sometimes in groups on both large and small projects.

Since it is so heterogeneous, it is difficult to define science precisely. The debate is aggravated when comparisons are made among sciences such as biology, mathematics, physics, sociology, history, archaeology, literary studies, political science, and gender studies. Are all these "scientific" in the same sense?

While sociologist Marcello Truzzi (1979) pointed out that how science is to be defined has been the focal point of many controversies, he distinguished between the social institution called "science" and the basic method of science. By accepting this distinction, scientific knowledge can be defined not so much by its content as by its form. That is, science can be thought of as the systematic recording of empirical data, organization of these data into logical categories, and the development of a plausible and persuasive explanation for these data.[7] Nevertheless, there are scientific disciplines (e.g., philosophy, mathematics) that do not require the collection of empirical data. R. G. A. Dolby, specializing in the history, philosophy, and sociology of science (1979, 10–11), suggested that "the most reliable immediate indication of what is scientific is provided by the mature judgment of the relevant expert scientists. . . . Orthodox science is that which commands the approval of all the leading scientific experts of the time." In other words, what is (and what is not) science is determined by the scientific elite. Consequently, a definition of something as scientific always involves a social evaluation. Harriet Zuckerman, a sociologist of science, chose to focus her definition on the normative structure in science. She distinguished between two classes of analytically separable norms: the cognitive (technical) norms and methodologies, which specify what should be studied, and the moral norms, which specify what the attitudes and behavior of scientists should be in relation to one another (Zuckerman 1977).

Scientific Research: Goals and Practice

It has become easier to discuss the goals of scientific research or science rather than its definition. Defining the goal of scientific research activity has indeed become an easier task since 1968, when Columbia University sociologist Robert Merton provided a characterization that seems to be acceptable to almost all scholars. Merton stated (1968b, 606) that "the institutional goal of science is the extension of certified knowledge" about the natural world. That is, scientific research aims to discover something true and valid about the world. This definition means that two important processes take place in such research: first, there is a social process that "certifies" knowledge; second, there is a cognitive process that aims to discover or create new knowledge. Producing or innovating new knowledge, however, can contradict the social process of certification; it can especially challenge old certified knowledge. Thus, for example, innovations in science may sometimes present an unconventional façade.

What Raises Research Puzzles?

Scientific explanations, be they historical, descriptive, functional, or experimental, share a common emotional psychological characteristic: they reduce, solve, or eliminate what was previously regarded as a riddle or puzzle (and sometimes replace what was considered a certainty). But cognitive questions remain: What is acceptable as a scientific explanation? What makes one explanation better or more persuasive than another? Why are researchers willing to accept or believe one explanation but not another? The philosophy of science has dealt extensively with these types of questions and has suggested numerous formulas.[8]

For our purposes it is important to note two things. First, while not all phenomena evoke puzzlement, turning one's attention to some so-called "obvious" phenomena could result in important discoveries and understandings. For example, pondering about an obvious phenomenon of why the sky is dark at night (also known as the "Olber's paradox") can clue us into understanding basic cosmological issues. The response is not because the sun is shining on the other side of the earth. The reason the night sky is dark could be either because there is not enough energy in the universe to light up the night sky, or because as the universe expands, light from distant galaxies is redshifted[9] and thus its contribution to the night sky brightness is diminished.[10] It thus appears that the most convincing modern solution to Olber's paradox lies in big-bang cosmology. The simple and obvious observation that the night sky is dark, which for most people does not arouse any perplexity, can lead to important and insightful ideas about the nature of the universe. Thus the type of questions we ask ourselves about our social and physical environment, and the type of events that cause wonder and curiosity, are anchored in our (social) definitions of reality. For researchers, it is the paradigm that reigns at the time.

Second, explanations that reduce a sense of wonder about something will or will not be accepted depending on the general worldview of the person or organization that developed that sense of wonder.

In a specific social system, a sense of wonder and awe was created by the sun, stars, and moon, and their movements were explained by accepting the theory that the Earth was the center of the universe. In the social atmosphere which prevailed during the late Middle Ages and the Renaissance, witchcraft, astrology, and alchemy were accepted as valid explanations for a variety of phenomena. Philosopher of science Thomas Kuhn offered the following rule of thumb for acceptance or rejection of explanations: "When reading the works of an important thinker, look first for the apparent absurdities in the text and ask yourself how a sensible person could have written them" (1972, xii).

What makes something more or less of an absurdity is, of course, determined by the personal or professional worldview of the person concerned. For many scientists, the theories of "ancient astronauts" fostered by author Erich von Däniken and others[11] constitute an "apparent absurdity." The same would be true of areas such as astrology, the cosmological theories of Immanuel Velikovsky, ESP, and pyramidology. During the nineteenth century in Germany, scientist Baron Carl von Reichenbach publicized the "phenomenon" of Od (after the Teutonic deity Odin).[12] Od was supposedly present in either positive or negative forms, and there were people who were supposedly sensitive to it and could see it flow from the fingertips. In the 1950s in the United States, psychoanalyst Wilhelm Reich suggested that living cells were made of "bions," pulsating units of "orgone" energy.[13]

Change in Science: Unconventionality vs. Deviance

The abundance of scholarly works in the areas of both the philosophy and the history of science reveals that the development of science demonstrates a much more discontinuous process than what is usually perceived (Truzzi 1979, 128). British historian Hugh Kearney's (1971) study of the scientific revolution in the sixteenth and seventeenth centuries demonstrated that it was not a slow and gradual process of change but a complex interaction among three different paradigms. The scientific process is so removed from the normative portrayal of it (i.e., gradual and accumulative) that historian Stephen G. Brush titled a 1974 paper in *Science* as "Should the History of Science be Rated X?"

Thus questions of what is accepted and what is rejected in science, where and when, may require complex responses. Conceptually, it is also important to distinguish between two distinct usages of "deviance" in research: (1) knowledge claims that we today consider absurd but that were presented and accepted sincerely (by some or many) at a different time, or vice versa; and (2) knowledge claims that were produced by intentional misrepresentation, that is, fraud.

One relevant issue here is sociologist Robert Merton's suggestion (1968a) that violating norms can sometimes be termed "innovation." Much more direct was sociologist Jack Douglas, who suggested in 1977 the term "creative deviance" arguing that "deviance is the mutation that is generally destructive of society, but it is also the only major source of creative adaptations of rules to new life situations" (60). More recently, sociologist Bernhard Giesen wrote, "Deviance . . . consists of turning the . . . evil moment . . . into a creative breakthrough to the unknown" (2005, 281). And, indeed, Wells et al. (2006) demonstrated empirically that deviance can be associated with creativeness, and Brisman's work (2010) coins the term "creative crime" to explore the role of such phenomena as cultural jamming, interventionism, and sabotage in inspiring cultural changes. The realization that both unconventionality and deviance can be easily interpreted as challenges to existing status quos does not mean that every criminal is a Robin Hood type of hero. It does imply that deviance can sometimes be a source of innovation and that given social conditions to pursue more or less freely one's creative ideas can open the door for destructive deviance. A sort of a diabolical equation may exist here; creating a space and atmosphere that promotes scientific creativity may also leave the door open to fraud.

Unconventional Ideas: Deviant Sciences. Polymath Michael Polanyi (1967) suggested that there are cases when experimental data should be dismissed when the data violates current scientific convictions. He further suggested that if the data relates to something real, the results would probably be produced again. If this occurs consistently over time, only then should change be considered.

It is difficult to evaluate the validity of Polanyi's argument since we do not have systematic data sets on the suppression of innovative ideas in science (Zuckerman 1977, 119). Sociologist of science Joseph Ben-David, however, did a study (1977) in which he reviewed discoveries that had at first been dismissed in the nineteenth century but later rediscovered and found to be significant. In his study, Ben-David states that the surprising factor was not in the prevalence of various attempts to suppress the ideas, but rather "the fact that they so rarely succeeded and that at no point was there any doubt among those with different prejudices that the contest of views could be resolved by accepted scientific procedures" (289). The only flaw in this argument is that there is a possibility that valid scientific conceptions and ideas were buried and never rediscovered.

Both Polanyi's philosophical conviction and Ben-David's empirical work cannot, by their very nature, account for either the number of cases of lost work that have never been recovered or their significance. For example, when in 1964 physicist, radio astronomer, and Nobel laureate Arno Allan Penzias and astronomer and Nobel laureate Robert Woodrow Wilson (and the Princeton team headed by physicist Robert Henry Dicke) discovered the most significant remnant of the primal big bang, the pioneering work done years earlier by physicist and cosmologist George Gamow and his group was—for all practical purposes—forgotten. Because Gamow and some of his original team were still living at the time, a controversy arose as to who should be credited with the idea. It is quite possible that many good, valid ideas have simply been forgotten because they were developed "before their time," that is, before the established scientific paradigm was ripe enough to absorb them. This is probably at a high personal cost to the original developers of the idea who felt that they had made a genuine discovery. This phenomenon is in total accordance with Truzzi's (1979, 131) observation that "like any form of deviance within a social group, unconventional ideas in science are seldom positively greeted by those benefitting from conformity."

While it appears that junior scholars aspiring to a secure academic position may need to be careful with new ideas, this is not the situation with established or marginal scholars. Ben-David (1960) illustrated how, in the fields of bacteriology and psychoanalysis, scholars who held marginal positions succeeded in introducing significant innovations. In the field of sociology, such innovations as the development of ethnomethodology and the sociology of the absurd did not originate from major, mainstream[14] figures. Both Max Heirich (1976) and sociologists Helga Nowotny and Hillary Rose (1979) demonstrated how "counter-culture" movements develop within science, creating the potential for both cultural/scientific and conceptual breakthroughs.

It is thus clear that the way in which routine scientific work is conducted involves sometimes unconventional knowledge claims. Some of these claims may become orthodoxy; others may be defined as deviant sciences (Ben-Yehuda 1985).

Misrepresentation and Fraud: Deviant Scientists. The other meaning of deviance in science refers to knowledge claims that were made by researchers who knowingly and deliberately violated scientific norms, for example reports of observations that have never been made. Scientific research proper may have little to do with such behavior, and such patterns of behavior were referred to in the past as "deviant" (Ben-Yehuda 1985, 1986a).

In perhaps a strange way, deviant sciences and deviant scientists are not completely separated. In both cases what we observe are creative, unconventional behaviors. In one case we have genuine attempts to introduce innovations that are rejected by the relevant community of researchers. In the other case we have deliberate attempts to falsify. One interesting hypothesis is that using mechanisms strong enough to stifle the second may have also adverse effects on the first. The next two sections in this chapter delve into these two lines of deviance.

DEVIANT KNOWLEDGE CLAIMS

While science aims to solve puzzles and innovate, it is nevertheless conservative and resistant to change. Kuhn (1962) suggests that "normal science" enjoys widespread acceptance by the particular scientific community that shares a specific theoretical and empirical paradigm. This paradigm provides guidelines for research methodologies and for prioritizing research goals, and it establishes criteria for accepting or rejecting data and hypotheses. Kuhn suggests that the history of science can be characterized by the dominance of specific paradigms. A scientific paradigm constitutes, for a longer or shorter period of time, the worldview, or the definition of reality, of specific scientific disciplines as well as the jargon used in that discipline. While a specific paradigm reigns, it will provide criteria for what is "sensible" and what is "nonsense." Phlogiston theory in chemistry, relativity in physics, Parsonian or ethnomethodological theory in sociology, psychoanalysis in psychology, and the idea that the earth is the center of the universe can all be used as examples of specific paradigms. Ultimately, the existence and persistence of anomalies will force an entire field into a state of crisis. This crisis situation generates questions relating to the basic assumptions, validity, and reliability of a specific paradigm. Kuhn stated that a crisis situation is solved by what he calls a scientific revolution,[15] which creates a new paradigm that will dominate the field until the next revolution occurs.

The Kuhnian conception provides a very different view from the one fostered by orthodox science. As empirical historical analyses indicate, however, the Kuhnian view is not always valid. Furthermore, Kuhn's account takes very little note (if any) of the *social* processes involved in a scientific revolution (Mulkay 1972a, 1972b). This is a serious omission because these processes are involved in significant ways in creating or disrupting relationships such as deceit, trust, and falsifications.

Some researchers working in the area of social movements have developed the concept of a "master frame" (and processes) attempting to conceptualize the ways in which the ideologies of various such movements are translated into the individual level. These "master frames" can be thought of as part of relevant paradigms or cultural cores,[16] that is, of symbolic structures that influence in the most profound manner the way we perceive, interpret, and act in the social environments in which we live.

Judging anything as "false" or even suspecting fraud needs to be understood within this context. For example, suspecting that a well-known research scientist is involved in fraudulent practices requires thinking not only about power relations but also about whether his or her work "makes sense." Thus these cultural cores are utilized by individuals in order to make sense of their empirical reality, at times transforming *factual truths* into *interpretive truths*. These cores help to produce criteria for, and sensitize individuals to, what to pay attention to, what to ignore, and how to integrate one's knowledge into a systematic, consistent, and coherent worldview. They provide us with important conceptual ideas that individuals acquire and that they use when they find it necessary to manufacture meaning, forge identities, and delineate social and moral boundaries. When any specific person relies on these cultural organizing principles, that person will accept and reject information as it fits those principles. An effect of deception can be easily created or cloaked when these cognitive processes are set into motion. Deception, however, requires more than just choosing selective information. This act of choosing must be made consciously, by deliberately suppressing or concealing other and possibly contradictory facts, and with the explicit intent to deceive unsuspecting audiences.

Factual and interpretive truth, as well as paradigms, master frames, or cultural cores, are crucial concepts when we try to understand how and why particular researchers allow their beliefs to impact the results of their studies. And indeed, we have little doubt that Blondlot, Kinsey, Lysenko, Cyril Burt, and others harbored a belief in the cultural cores within which their paradigmatic scientific worldview was embedded and that helped organize their consciousness about how the world ought to be.

No need to conclude such a discussion without noting sociologist and political scientist Richard Hamilton's intriguing work (1996). He analyzed how some major and favorable social theories were accepted despite accumulating evidence against them. Indeed, Hamilton points out that even in social science, supposedly based on the search for and presentation of empirical evidence, truth does not always prevail.

The important points for our discussion are not only Kuhn's view of change in science, but the implication of his work that science is conservative and resistant to change. There are numerous illustrations for this conservative stand. One illustration is the standard hypothesis testing using statistics when 95–99 percent probability is given to existing knowledge (the H_0 hypothesis claiming that the known knowledge is valid). Another illustration is the process through which papers and books (or grant applications) are accepted for publication by what are considered to be prestigious and reliable journals and publishers. In such cases, manuscripts have to be reviewed by usually three (and sometimes more) reviewers. This procedure can be thought of as not only a good safeguard against the publication of nonsense, but it is also an excellent mechanism for maintaining stability. This procedure, obviously, makes it important who the reviewers are, and thus editors have much power in helping a specific report to be published (or preventing it) simply by deciding who the reviewer/referee would be.

In addition to the two examples mentioned above, the concept of professional technical competence (i.e., the recognition that the credibility of a certain individual's work is reinforced by years of accumulated experience in the field) creates sets of vested interests and controversies over the nature of a phenomenon under study. For example, cytogeneticist Barbara McClintock, recipient of the Nobel Prize in medicine in 1983 for her work on genetics, was clear in interviews to the electronic and print media[17] that in the 1950s her colleagues called her crazy, that she could not find a job, and that she was ridiculed. Her work on the so-called "jumping genes" ("transposition," see Keller 1983, chap. 8) was certainly considered deviant (Lewin 1983; Judson 1979, 460–61). It took the scientific community about thirty years to acknowledge that McClintock was not—after all—crazy. Brush (1974, 1167) succinctly summarizes this point, quoting the famous scientist Max Planck's bitter remark about the fact that new theories "rarely get accepted by rational persuasion" and that "one simply has to wait until the opponents die out." Gould (1977, 160) noted that "as the new Darwinian orthodoxy swept through Europe, its most brilliant opponent, the aging embryologist Karl Ernst von Baer remarked with bitter irony that every triumphant theory passes through the following three stages: first it is dismissed as untrue; then it is rejected as contrary to religion; finally it is accepted as dogma and each scientist claims that he had long appreciated its truth."

Thus resistance to change and innovation in science has a few sources. Disciplinary paradigms and peer reviews (Schlussel 1983) are indeed such major sources. A 2015 instructive study into this issue is provided by Dreger's fascinating work about the nature of resistance and truth in science, and especially her exposure of liberal politicization of science.[18]

This state of affairs explains why it is so difficult and risky to introduce new facts or new interpretations. While a well-established, prestigious scientist can afford to take risks, others who do not enjoy that status can be expected to exhibit a more restrained and risk-avoiding behavior. Young scientists in particular are faced with a dilemma of whether to take the risk or to avoid it. Especially for those seeking academic careers, taking risks can be dangerous if they do not have prestigious scientists to back them up. On the other hand, a successful risk can boost a young scientist's career tremendously.

The field of quantum mechanics, for example, has become one of the oddest areas of research, at least for outsiders who are not directly involved in active research. Some of the research's conclusions and theorems can indeed seem bizarre to the layman (e.g., application of Bell's theorem to quantum mechanics). Social scientist Bill Harvey (1981) showed how various knowledge claims were evaluated in experimental tests of quantum mechanics. Harvey showed how, in each case, an agreed-upon evaluation became possible only due to the shared culture of the physicists involved, making certain evaluations plausible and others not. Furthermore, Brush (1980), who researched the philosophy of quantum mechanics from a historical perspective, showed how difficult and full of objections the path of quantum mechanics was until it became established, accepted, and legitimized.

Research in parapsychology faced a similar problem, for one of the major hurdles for parapsychological research has always been to establish, preferably in controlled experiments, the reality of parapsychological effects. The difficulties in accomplishing this task have been documented (i.e., Mauskopf and McVaugh 1979) and have given rise to heated arguments as to whether parapsychology is science, pseudo-science, or, perhaps, proto-science.[19] Sociologist Trevor J. Pinch (1979) demonstrated, among other things, how a hypothesis of fraud (i.e., the assumption that something must be wrong in that area of research) was applied to parapsychological research, causing the rejection of plausible explanations of parapsychological effects. The fraud hypothesis denies the very reality of parapsychological effects.[20]

The negation of specific aspects of reality in the natural sciences need not be confined to esoteric and marginal activities such as parapsychological research. The debate between creationists and evolutionists provides another vivid example of challenging a specific construction of reality. Negotiations and arguments about the nature of reality have obviously not been confined to the natural sciences only. Theoretical approaches in sociology such as ethnomethodology (Garfinkel 1967) and the sociology of the absurd (Lyman and Scott 1970) exemplify this. Furthermore, the dramaturgic approach to the analysis and interpretation of social interactions, as fostered by Erving Goffman and by symbolic interaction as well, demonstrate that reality cannot only be negotiated but also that the claim that there really is "something out there" can sometimes be problematic at best. Phenomenological research in deviance, exemplified by sociologist Jack Douglas's (1967) work on suicide and sociologist Alan Blum's (1970) work on mental illness, illustrates this claim. Even the more eclectic interpretations in sociology and psychology, such as those made by Karl Marx and Sigmund Freud respectively, also focused on particular reality constructions as well, hinting perhaps at "ultimate truths" (not to mention works in so-called "speculative history" such as those by Toynbee). Such essentialist approaches obviously take various reality constructions as not problematic.

This discussion takes us back full circle to basic questions that were posed earlier and that focused on the nature of a particular explanation and on reasons why certain explanations were rejected while others were accepted. Scientific explanations must make some assumptions about the very nature of reality. Astronomer Fred Hoyle's revised version of his original steady state cosmology is based on different assumptions regarding the nature of reality than those held by proponents of the big bang and expanding universe cosmologies. Those adhering to sociologist Talcott Parsons' assumptions (and view of society), which hold social reality to be stable, obvious, and noncontroversial to the trained observer, would grasp sociological processes and society quite differently from those adhering to ethnomethodological views or conflict theorists.

Scientific disciplines, therefore, constitute a complex matrix of beliefs and are characterized by professional and ethical cultural cores. This matrix includes the prevailing disciplinary paradigm and the scientists' criteria for evaluating knowledge claims. "Plausibility" and "explanation" are therefore defined within this matrix of existing beliefs (Barnes 1977, 279).

This situation framed what can be easily called deviant sciences. We suggest that such sciences are knowledge claims that, by virtue of their hypotheses and/or methodology, challenge scientific cultural cores and are consequently regarded by the relevant scientific establishment as deviant. Hence these claims are viewed by orthodox or mainstream research scientists as suspect.

In this context, three main concepts can be distinguished. One is protoscience;[21] that is, knowledge claims that aspire to become mainstream science and seem to be on their way to doing so (psychoanalysis was, for a while, in that position). The other is pseudo-science (sometimes referred to as junk science,[22] voodoo science, or other such "complimentary" labels); these are knowledge claims that are regarded by mainstream scientists as problematic, suspect, and questionable. UFOlogy is a good illustration. The third category is that of fringe or speculative science. These are typically knowledge claims that are regarded as on the periphery of science. Continental drift theory or radio astronomy once occupied that position. For our purposes, we prefer using the more general name "deviant sciences." Bearers of these knowledge claims usually have a statement to make about what they view as the "true" nature of the world. Thus they challenge orthodox science regarding the very nature of the scientific endeavor. In principle, such claims can shatter conventional theory or methodology or both. Naturally, we can expect the relevant scientific community to take such attempts seriously and fight them. This resistance can assume many forms.

Deviant sciences challenge the rationality (and consequently, its boundaries) of orthodox science and may often try to defeat scientific monopoly regarding definitions and explanations. Typically, an essential element in these challenges is an explicit or implicit argument about the nature of reality. Do UFOs actually exist? Does the Loch Ness monster exist? Is it plausible to associate quantum mechanics with the psychological concept of free will or of consciousness? Is there a "grand design"? Did ancient astronauts interfere with our cultures? Can one use thoughts to read other people's minds or move objects?

And yet, defining "deviant science" is not a simple undertaking (Nowotny and Rose 1979) because demarcating "science" from "nonscience" is not always an easy task (Gieryn 1983). Mostly we can claim that deviant sciences challenge regular scientific disciplines and, in particular, the paradigms (in the Kuhnian terminology) of normal science. The term *deviant science* thus refers to "strange" observations, to anomalies that are not recognized as such by normal science, and to bizarre interpretations about the nature of reality. A deviant science's claims may threaten paradigms of normal science, which partially explains why deviant sciences are frequently so fiercely rejected and stigmatized by the science establishment. This can be observed by the responses of disbelief, ridicule, rejection, isolation, or stigmatization by orthodox or mainstream science. The reactions to deviant sciences, and those adhering to them, are similar to reactions to regular deviance and include rejection, isolation, stigmatization, ridicule, and expulsion from jobs. The case of Velikovsky serves as a good example for this.

A few scholars have written about deviant sciences. The first formulation was made by Irving Langmuir (1953), a Nobel-winning chemist, who coined the phrase "pathological science." Pathological science[23] is not quite deviant science because the term Langmuir suggested referred primarily to "that which cannot be," that is, to reports about observations that were never made or, if you like, to fraud. Deviant sciences, in many cases, are not like this.

Two more formulations appeared almost simultaneously in 1979. The first was a book edited by sociologist Roy Wallis and titled *On the Margins of Science: The Social Construction of Rejected Knowledge*. The first article, written by Dolby, deals extensively with deviant sciences. The other chapters deal with subjects such as the normalization of deviance in science, astrology, mesmerism, acupuncture, phrenology, creation vs. evolution, spiritualism, paranormal claims and parapsychology, sea serpents, and UFOlogy.

Dolby's chapter on deviant science is obviously of special interest to us. Dolby suggested (10) that "[o]rthodox science is that which commands the approval of all the leading scientific experts of the time." In contradistinction, deviant science (11) "is that which is rejected by the orthodox scientific experts, and which they may label 'pseudo science'. . . ." Most of Dolby's (1979) paper is indeed devoted to an analysis of belief systems that are in conflict with orthodox science. Thus what constitutes scientific and nonscientific may sometimes become a matter of social evaluation and recognition. Specializing in the study of science, Brian Wynne's (1979) analysis of the J-phenomenon supported this claim, showing how deviance in physics was normalized and pointing out that there are no absolute boundaries between scientific controversies and encounters between orthodox and deviant sciences. Likewise, author Eugene Frankel (1976) demonstrated how the adoption of the wave theory of light over the corpuscular system of the early nineteenth century drastically altered the field of optics and posed the first serious difficulty for the action-at-a-distance worldview that had dominated European physical science from the time of Newton (141). While Frankel vividly described the conceptual breakthrough introduced by the idea of "light as waves," he also demonstrated how difficult it was to initiate this change and how orthodox science rejected the concept. Only after optical physicists lost some of their best colleagues to the wave theory did they stop ignoring it and allow change to be gradually introduced.

The social position of those adhering to various deviant sciences, especially within the scientific community, is of crucial importance if we wish to understand why a "deviant" suggestion becomes legitimate (that is, a case of proto-science) and vice versa (that is, a case of obsolete science). To a large extent, power plays a crucial role here. Sociologist John Lofland (1969, 14) characterized deviance as "the name of the conflict game in which individuals or loosely organized small groups with little power are strongly feared by a well-organized, sizeable minority or majority who have a large amount of power." Sociologists Horowitz and Liebowitz reinforced this definition when they stated, "Deviance is a conflict between at least two parties: superordinates who make and enforce rules, and subordinates whose behavior violates those rules" (1968, 181).

Deviant sciences generally fit these characterizations. In most cases we have an individual or a small group of scholars or researchers whose challenge may raise some anxieties in a sizeable group of fellow scholars. Indeed, the reality and specific scientific claims made by the minorities can potentially challenge the agreed-upon paradigms of the dominant normal or orthodox science. The Velikovsky affair illustrates this point. In brief, Velikovsky suggested that various religious myths had a solid basis in reality. He claimed that there were physical upheavals of global proportions in historical times and that these upheavals were caused by extraterrestrial reasons (mostly changes in the structure of the solar system). Sociologist Robert Mcaulay (1978) points out that when Velikovsky presented his ideas to Harlow Shapley (an eminent contemporary astronomer) in 1946, Shapley responded angrily by writing letters to his colleagues stating that "if Dr. Velikovsky is right, the rest of us are crazy." (317). Later a campaign against Velikovsky was organized in which Shapley threatened the Macmillan Publishing Company and asked them not to publish any of Velikovsky's books. In 1952, the American Philosophical Society gave an opportunity to attack Velikovsky[24] but did not allow him to publish a rebuttal. These two examples only demonstrate the type of organized fierce resistance Velikovsky encountered.[25] Mcaulay (1978) suggests that organized "opposition to Velikovsky among segments of the scientific community was partially mobilized along informal lines by mail, direct contact and, apparently, the personal authority of Harlow Shapley."

Deviant knowledge claims may generate a controversy in which the status of the researchers who suggest or support the deviant claim may help to determine whether it will be accepted or rejected. Among other things, the integrity, past achievements, and professional credibility of those who support and those who reject may become an issue as well. Sociologist Howard Becker's (1967) theoretical concept of hierarchy of credibility is useful here. According to Becker, societal hierarchies of credibility structure interactions whereby some people are considered to be more credible and reliable than others. The more secure and prestigious a researcher's position in the scientific community is, the more credibility he or she enjoys.[26]

Processes of change in science do not resemble a democratic procedure. Deviant sciences must inevitably face scientific and political struggles for recognition and legitimization by relevant researchers in the scientific community. These struggles can be fierce—sometimes even nasty—because the challenge debated in them sometimes concerns the very core of certified knowledge, its methodology, and sometimes even the worldview behind it and its rationality. For example, accepting knowledge claims of adherents to ancient astronauts' theories implies a very different understanding of the development of humans and their culture and history.

UFO claims have been discredited and stigmatized throughout by both the normal scientific orthodoxy and official government bodies (at least in the United States). One of the results of such discreditation and stigmatization (especially if adherents to the deviant science are convinced and committed to their knowledge claims) is the creation of organizations that promote and support the deviant knowledge claims in question outside established and recognized research frameworks and that may gain significant popularity outside the usual networks of academic researchers. Astrology is a good example. Proponents of such subjects as spiritualism, ESP, and Fortean[27] events have all been successful—at one point or another—in establishing organizations and institutes that promote these knowledge claims, present them as "sciences," and whose ranks are filled with enthusiasts. In some cases, the extra-academic activity may indicate a high level of popularization (e.g., Däniken and the followers of the "ancient astronaut hypothesis," UFOs, Velikovskian ideas, including popularized TV channels such as National Geographic, History Channel, and H2[28]); in other cases it can create the impression of a quasi-academic activity.[29] Adhering to deviant sciences that are usually stigmatized and rejected can also give rise to cults. This was evident in the UFO case (Story 1980, 506,

305–8) and in groups such as the internationally famous TM and Scientology, or the more local (e.g., British and Israeli EMIN).

Some deviant sciences can be referred to as modern myths (Goran 1980) and were part of the occult or new age revivals (Ben-Yehuda 1986b). Modern myth can consist of Däniken's famous theories about "ancient astronauts," theories about "lost continents," the Bermuda Triangle mystery, pyramid power, UFOlogy, the Loch Ness monster (and other monsters, such as Bigfoot), to mention only a few. These deviant sciences/modern myths attract hundreds of thousands—possibly even millions—of believers. Similar to sociologist Andrew Greeley's (1970a, 1970b) experience in the early 1970s, we may face advanced undergraduate and graduate students who may find it difficult to accept that the ancient astronaut theory, for example, lacks even one, irrefutable, replicable hard fact to support it. Those very students who may spend hours trying to refute Parsonian or capitalist theories would not consider spending even one percent of that effort in an attempt to critically examine Däniken's theories or the way they are presented. Indeed, psychologists Barry Singer and Victor Benassi (1981) stated that the lack of enough science education is to be blamed for this phenomenon.

Sociologist Michael Carroll (1977, 548) suggested an analysis of modern myths through Levi-Straussian structuralism because "like those living in primitive cultures . . . concerned with the resolution of certain universal oppositions . . . the popularity of the modern Atlantis myth and the myth of the ancient astronauts derives from their ability to effectively resolve these oppositions." Another Levi-Straussian analysis—considerably different from Carroll's—was suggested by sociologist C. E. Ashworth (1980, 374). He notes that "Religion and Science . . . no longer resolve certain structural contradictions which they once did with ease," while modern myths do seem to solve these structural contradictions.

Deviant sciences do not necessarily change normal science, and they can remain deviant. Seymour Mauskopf's edited volume (1979) reflects a symposium sponsored by the American Academy for the Advancement of Science on the topic of the reception of unconventional science.[30] The book provides detailed analyses of four cases of what were once considered "unconventional" sciences but which later received acceptance: acausal quantum mechanics, Wegener's continental drift theory, acupuncture, and parapsychological research. The book concludes with a chapter by Truzzi that suggests that deviant science (or pseudo-science) is one that is methodologically flawed. Truzzi suggests the term "proto-science" to describe knowledge claims that assume the existence of an extraordinary variable (e.g., UFO, unicorn); and finally "para-science" to describe claims that assume the existence of extraordinary relationships between ordinary variables (e.g., astrology).

One of the results of this situation is that we are constantly being flooded with books, papers, and media programs assailing us with all sorts of strange and supposedly unexplained phenomena, demonstrating how Western positivist-oriented science is at a loss to explain these prima facie strange occurrences. The academic community does not remain silent, and a counterattack may sometimes take place. Numerous books have been published that attempt to expose various deviant knowledge claims such as UFOlogy, ESP, parapsychology, Velikovskian-oriented claims, astrology, and rhythm of life.[31]

The relevant issue for us is the need to make a distinction between deviant scientists—those involved in fraudulent practices—and deviant sciences, which are not involved in fraudulent practices but try to change the course of science for reasons that are not antagonistic, malevolent, or malicious. That there may be cases of misconduct in deviant sciences is obvious. Nevertheless, our focus will remain on deviant researchers and not on what mainstream science refers to as deviant knowledge claims or those who promote them; that is, on researchers doing normal science in reputable research institutions but resort to fraudulent practices. It is instructive that some authors felt that the discussion of deviant sciences needs to come together with that of fraudulent sciences. For example, physicist Robert Park's book (2000) explores what he refers to as "voodoo science," which consists of genuinely foolish enterprises to blatant frauds. British journalist David Aaronovitch used the title "Voodoo Histories" to discuss how conspiracy theories molded perceptions of history.[32] Also relevant is historian Ronald Fritze's (2009) work that focuses on "invented knowledge" going from facts to speculative ideas in the sciences, history, and religion.

DECEIT: SOME ILLUSTRATIONS

Using such captivating terms as "lies," "deceptions," "con games," "cheating," "scheming," assumes not only that some establishable "truth" exists, but that it is subverted. To illuminate some of the issues involved, as well as to whet our appetite and curiosity, let us examine in more depth a few famous illustrative cases from the annals of deviance and deceptions in scientific research.

The case involving Mark Spector. At the annual meeting on tumor viruses at Cold Spring Laboratory in May 1980, Mark Spector, a twenty-four-year-old graduate student in biochemistry from Cornell University, presented a revolutionary paper. Spector suggested a new explanation of how viruses work and cause cancer. Steven O'Neal, a postdoctoral researcher in Cornell's biochemical laboratory, had worked unsuccessfully for more than two years to isolate a specific protein (sodium-potassium ATPase). Spector came to Cornell in December 1979 and was assigned the task of isolating the protein. He accomplished it in less than a month. This discovery was the first milestone in Spector's development of a new, innovative theory about viruses and cancer. The news of his findings spread rapidly in the scientific community, and in July 1980 some of his work was published in respected journals such as *Science* and *Cell*.

In the fall of 1980, however, signs of fraud were already apparent. At that time, Raymond Erikson, a virologist at the University of Colorado Medical Center in Denver, found that a sample serum from Spector's experiments contained certain antibodies that, if the theory was correct, should not and could not be present. Another warning sign was that a few of the experiments worked only under Spector's very personal supervision. By late July, some of Spector's own laboratory colleagues began to be suspicious. During the summer of 1980 Spector was asked to stay away from the laboratory while his supervisor, biochemist Prof. Efraim Racker, tried to replicate Spector's experiments. New discrepancies appeared. In September 1980 Racker sent letters to both *Science* and *Cell* partly retracting the previous reports, and he also withdrew papers that were scheduled to appear in two other journals. The university began to investigate Spector's past and discovered that he had never received a BA or MA degree. Spector withdrew from Cornell and denied all the allegations. His claim seems to be supported by the puzzling fact that some of his work does indeed seem to be valid and replicable.[33]

The case involving Vijay Soman. The Soman case (Broad 1980b, 1980c) occurred in 1980 at the Yale Medical School when Dr. Helena Wachslicht-Rodbard, a young endocrinology researcher at the National Institutes of Health, claimed that medical researchers Philip Felig and Soman, in one of their papers, had plagiarized some of the words and ideas of a paper of hers that they had seen that had not yet been published. She also wrote to the dean of the Yale School of Medicine and questioned the "authenticity" of the data in their paper and claimed that they did not conduct the reported research.[34] This case also received much contemporary publicity, as it appeared in the November 1, 1981, issue of the *New York Times Magazine*. Following a scientific audit, Soman—a young researcher in desperate need of publications—admitted to having falsified work in a paper. After an investigation, twelve papers involving Soman as a researcher were retracted (Broad 1982, 479).

The case involving Prosper Blondlot. This case concerns the "discovery" of N-rays.[35] The reputable French physicist from the University of Nancy, Prosper Blondlot, claimed that he had detected N-rays. This claim was made in 1903, about eight years after German physicist Wilhelm Konrad Roentgen discovered X-rays (for which he received the first Nobel Prize in physics in 1901). The period was rife with all sorts of discoveries focused on rays, and Blondlot had apparently pressured himself to discover some new form of ray. This "discovery" was well-received in France, and the French Academy even awarded Blondlot a prize, but American physicist Robert Wood was skeptical. He visited Blondlot in his French laboratory and while "Blondlot was observing and describing an N-ray spectrum, Wood slyly removed an essential prism from the apparatus. This had no effect on what poor Blondlot fancied he was seeing" (Gardner 1952, 345). As Asimov (1979) stated, Blondlot "wanted to believe in something desperately—and he did" (137).

The case involving Trofim Lysenko. As science writer Martin Gardner (1952, 149–51) noted, "Seldom in the history of modern science has a crackpot achieved the eminence, adulation, and power achieved by Trofim D. Lysenko, the Soviet Union's leading authority on evolution and heredity. Not only have his opinions been pronounced dogma by the Kremlin, but his Russian opponents (whose views are held everywhere but in the USSR) . . . have been systematically eliminated from their posts" (140). Lysenko, a former peasant and plant breeder, publicly defended and endorsed the Lamarckian view at a time when this view was debunked.[36] It stated that the mechanism by which evolution works is the inheritance of traits various organisms acquired when responding to their environment: inheritance of acquired characteristics. Lysenko's views accorded fully with the political views of the Soviet Communist Party at that time. His view supported contemporary revolutionary Soviet ideas about creating new Soviet men and women. Inheritance of acquired traits meant that once the first generation acquires certain traits (in this case Soviet ideas), the next generations will have those traits. The year 1948 marked a decisive victory for Lysenko and the agonizing end of his opponents, whose views (that were scientifically valid) were discredited by political motivation. No one needed to go too far to realize what a fraud Lysenkoism[37] was. One well-known and simple fact at the time that contradicted the view proposed by Lysenko is circumcision. It is clearly an acquired trait and it does not pass from generation to generation.

Like the case of Blondlot and N-rays, Lysenko might have believed—or wanted to believe—in the inheritance of acquired traits. Such a belief could go well with the communist attempt to create a new Soviet persona. Nevertheless, sticking to proper scientific procedure would have eliminated this fantasy. Although some new thoughts and developments may make some of Lamarckian views perhaps less objectionable,[38] that was not the case when Lysenko's opponents were terrorized. Perhaps one difference between the case of Blondlot and that of Lysenko is that the last was endorsed by organs of a state. Nevertheless, the prize Blondlot received from the French Academy could be interpreted as state-institutional support. For us, the important point is that in all these cases the researchers ignored—repeatedly—proper scientific procedures in their attempts to substantiate a belief. We do not believe that they committed honest mistakes, and there is no reason to exclude such cases from the designation of fraud.

The case involving Alfred Kinsey. As many of us may remember, the 1948 publication of biologist and sexologist Alfred C. Kinsey's book on the sexual behavior of males (better known as the "Kinsey Report"), and of another book on the sexual behavior of females in 1953, helped bring a revolution into being. The 1948 work by the Indiana University professor indicated that a large proportion of males were engaged in masturbating and premarital sex and utilized the sexual services of prostitutes. It showed that close to 40 percent had experienced homosexual contacts leading to orgasms, and that 30 to 45 percent of married males had extramarital sex. Similarly sensational findings were produced in 1953 about females. Within a slightly puritanical social context, Kinsey's books came as a shock. Basically, his research indicated that the sexual behavior of both males and females presented significantly richer, more varied and complex patterns than those allowed (or thought to be existing) by prevailing conservative morality. Kinsey, of course, was viewed as a dry, detached, and pedantic academic whose data collection was not something to be questioned on moral grounds.

Author James H. Jones' 1997 works casts dark shadows on this. He claims that

> Kinsey . . . was not quite what he appears to be—the genial academic in baggy tweeds and bow tie, the simple empiricist disinterestedly reporting his data. . . . [H]e was, in reality, a covert crusader who was determined to use science to free American society from what he saw as the crippling legacy of Victorian repression. And he was a strong-willed patriarch who created around himself a kind of utopian community in which sexual experimentation was encouraged. . . . In his obsessive energies and powers of persuasion, Kinsey resembled a late twentieth-century cult leader . . . with a burning belief in his mission. (1997a, 100)

Jones' work reveals the considerable disparity between the private and public lives and presentations of Kinsey. In public, Kinsey had the presentation of a cold, calculated, and careful scientist. In private, Kinsey was a domineering man, interested in and mastering many areas (entomology, biology, education, classical music, designing a house and elaborate flower garden), but above all he had a remarkably free and wild sexual lifestyle. It included experimentation with various sexual behaviors—homosexuality and masochism, promiscuity—all the while maintaining his role as (heterosexual) husband and father. Much of the sexual experimentation, by the way, was justified as "needed for research." Jones has no hesitation in concluding that Kinsey's private life and morality almost certainly affected the objectivity and detachment of his work as a scientist, and his celebrated findings may have been skewed. He points out that from the beginning of Kinsey's research into sexual behavior, the Americans who most persistently engaged Kinsey's attention were people who were either on the margins or beyond the pale: homosexuals, sadomasochists, voyeurs, exhibitionists, pedophiles, transsexuals, transvestites, fetishists (1997b, 101).

Jones notes that Kinsey was involved in outright deception when he had to justify how and why his interest in human sexuality developed. His presentations always gave the impression that he became involved in the topic quite by accident and in response to demands from his students. That presentation both concealed and falsified the truth, which was that Kinsey had been involved in sexual experimentation throughout his entire adult life, that some of his experiments were considered by contemporaries as esoteric, marginal, and negative, and that his private life certainly affected both how and why he became involved in sexual research (1997a, 105). The implication is that Kinsey was motivated to show American society as extremely unconventional in sexual practice but very condemnatory in principle, and therefore hypocritical. His crusade was to eliminate this hypocrisy, and to this end he used his scientific status and credibility.

Although the media's attention to Kinsey's work seems to have been generally favorable, the academic professional reaction was mixed. While no one knew about the formidable disparity between the public appearances Kinsey gave and the private realities he was involved in, his sampling methods drew sharp criticism, his book was described as confusing, and he was accused (by no less than Margaret Mead) of decontextualizing sexual acts from relevant settings.

Researchers who focus only on facts that support their research hypotheses and choose to ignore contradicting relevant facts run a risk of conducting a seriously flawed project.[39] When this practice is not a result of ignorance or mistakes but is meant to support some of the researcher's views, then obviously a deliberate fraud is probably involved, the main goal of which is to present a misleading representation of reality. Thus even if the carefully and biased selected facts presented by the researcher are valid, the overall result of the research is invalid.[40]

Three more famous cases we shall note in passing only involve Dutch psychologists from Amsterdam: Diederik Stapel, Harvard's Marc Hauser, and Britain's Edward Erin. The case involving Diederik Stapel is more recent and created global media waves, but it is—relatively speaking—a simple and straightforward case. In essence, Stapel—a social psychologist—was suspended in 2011 from his Tilburg University positions after being suspected of using faked data and of falsifying and manipulating at least fifty-five papers.[41]

During 2011–12, Marc D. Hauser, a Harvard University professor in evolutionary biology and a leader in his field, was forced to resign after suspicions of misconduct. The investigation into his misconduct began in 2007 and was later continued by the ORI. It was found that he "had fabricated data in one study, manipulated results in multiple experiments, and incorrectly described how studies were conducted." Interestingly enough, and typical of too many cases like this, the Hauser case created some controversies and drew attention. One issue was that "[t]he details of this investigation were not publicly released (which is common for most universities, including Harvard), and the lack of transparency evoked substantial speculation."[42]

Finally, it is well worth noting that falsified research may end up with tragic results. In June 2013 the British newspaper *The Independent* informed its readers that Dr. Edward M. Erin fabricated his work on asthma tests and that this fabrication ended up with some fatal results.[43] It was suggested in the newspaper report that, perhaps driven by a desire to enhance his career, Dr. Erin had consistently changed results for about three years. Dr. Erin tried to poison his girlfriend and was found guilty and sent to prison. The judge—Judge Richard Hone, QC—characterized him as an "'egocentric' who had delusions of grandeur" and as a "liar, a cheat and a predator." One of the scientists involved in the study, Peter Barnes, told the newspaper that Dr. Erin was "a liar and a cheat. But psychopaths are very convincing. It could happen to anyone."

FRAUD IN OTHER CONTEXTS

Author Jeremy Campbell (2001) argues that lies and falsehoods are a natural—not deviant—aspect of human behavior. Accounts about deceptions, falsifications, and hoaxes are easily found in many other areas than scientific research. Let us have a brief look at some of these areas. It appears that, overall, hoaxes are probably leading this complex human behavior that involves untruth.[44] While book collections[45] about hoaxes cover such conduct from various fields, they also treat some fraud in research as hoaxes.[46] Other cases are found as media hoaxes,[47] archaeological,[48] and geographical exploration hoaxes (Roberts 1982). Some works are focused on one specific hoax, for example (a hoax in science) the Sokal hoax (2000),[49] or the Gospel hoax (Carlson 2005). Close to hoaxing, we find published cases of frauds and deceptions,[50] among them fabricating a land (Sinclair 2004), or mining (Plazak 2006). Fakes and forgeries in the arts have also attracted much public attention,[51] as have the histories of some major fascinating cases of deceptions,[52] falsehoods (Campbell 2001), and fraud in history.[53] Some frauds involving Ponzi schemes drew much public attention with the case involving billion-dollar swindler Bernard Lawrence Madoff reaching new heights in 2008–2009. Other swindling schemes drew attention due to their dramatically outrageous nature. For example, Victor Lustig's 1925 attempt to sell the Eiffel tower (twice).[54] Naturally, some authors wondered why we tend to believe some strange tales, and what the nature of these tales is.[55] It is thus important to keep in mind that cases of fraud and deception in research are part of a larger category of similar acts practiced in other areas as well.

Lies and deceptions thus occur between and among individuals, groups, and nations, ranging from the unclear and problematic to the clear and obvious. Thus infidelity, military deception, national lying, faking orgasms, computer crime, embezzlement, insurance fraud, "con" games, worthless medications, plagiarism, environmental crimes, consumer fraud, counterfeiting money, fashion products, music CDs, and art, and a host of similar phenomena are included under the umbrella of "fraud" and "deceptions." The nature of this range is its kaleidoscopic variance and its dichotomous and continuous variables. Moreover, our language allows us to make fine distinctions among different lies and deceptions, for example, "blatant lies," "deceptions," "white lies," or "exaggerations." The specific term we choose to use in specific instances of "lies" is context-dependent and hinges on how we want to construct an interpretation for a specific case. Issues pertaining to the nature of the lie, its impact, the intentions of the liar, the audience, all are variables that help us decide, case by case, how to classify the tale: innocent lie, exaggeration, or blatant lie.

We live in a cultural ecology that is constituted of, and incorporates, lies and deceptions into its very essence. Lies and deceptions are therefore to be found almost everywhere and any time. The crucial variable in this context is one of degree and quality, not of principle. The fact is that since various deceptions are part and parcel of our cultures, we are necessarily involved in at least some of these deceptive schemes. Consequently one of the issues we need to examine is the very nature and characteristics of lies. Academic interest in lies and deceptions has not been too strong, but it seems to have gained momentum, which has yielded some interesting formulations.[56]

Society, Culture, and Lies

The fact that our cultures have lies as an integral part means that we, humans, have the capacity to use symbolic means of communications to transcend the limits of time, space, and geography. There can be little doubt that the ability to fantasize is intimately associated with the ability to lie and deceive, and that these capabilities are greatly facilitated by the acquisition of language skills, oral and written. Thus we are capable of creating legends, myths, and fairy tales, we can fabricate, cheat, deceive, and lie; in other words, we can relate to things whose empirical reality is partial, imaginary, nonexistent, and deliberately falsified. Researchers of lies tell us that we all tell different types of lies and are involved in various deceptions. These include defensive lies, political lies, romantic lies, and so on. The list of types of lies that educational expert Adir Cohen constructed (1999, 124) is indeed impressively long. Social psychologist W. Peter Robinson even states explicitly that "lies have great potential for preserving the status quo" (1996, 299). These researchers also note that the construction of truth is meaningful only when we contrast it with "lies and falsehood." In fact, there are social roles in which lies are integrated as a matter of course. For example, there are diplomats who are engaged in fabricating deliberate lies in an explicit attempt to conceal the truth; they are expected to lie and to repeat those lies in public; it is part of the profession, we are told. Another example, perhaps a more positive and instructive one, is the so-called Witness Security Program in the United States (and in other countries). This program involves official bureaucratic agencies in a coordinated activity whose aim is to falsify identities and keep those falsified identities concealed for as long as possible.

As the previous example indicates, not every act of telling a deliberate lie is considered culturally negative. In fact, some such activities are not usually even thought of as telling lies. "Storytelling"—in the form of creating fictional narratives, characters, plots, and places; movies and other forms of art that are loosely connected (if at all) to reality as we recognize it—can be generously rewarded. The global interest in the Harry Potter and Star Wars tales is a good illustration. Obviously, one of the major differences between Tom Clancy's or Alistair MacLean's fictional suspense and fantasy war stories and criminal "con" artists is the intention. Creating fiction for the sake of fiction, deliberately but openly, can be a socially and economically gratifying activity. The positive potential of telling such fables is that they can entertain, teach, be a guide to memory, and explore difficult and complex issues that readers may face in real life. But telling deliberate lies that conceal the truth and manipulate unsuspecting victims in a way that is detrimental to their well-being is typically condemned as negative and deserving of punishment. Hence the activity of telling lies, in itself, is not universally forbidden. On the contrary: It is the context of that activity that will help us determine what the appropriate reaction to telling these lies will be. In the illustrations we used, intention seems to have been the crucial factor.

Deceptive lies can indeed sometimes be considered quite useful. Using deception in war or intelligence is an ancient practice. One of the better-known examples is the horse left by the Greeks for the besieged city of Troy. More recent examples involve tricks used by the Allies during World War II such as General Patton's phony eleven division's army, the First United States Army Group (FUSAG). In fact this military "force" never really existed and its creation was a deception aimed to mislead Hitler as to the landing of Allied forces in Europe. Another famous deception was Operation Mincemeat's use of a corpse to help mislead the Nazis as to where the Allied landing in the Mediterranean was supposed to take place (Macintyre 2010; Smyth 2010).[57] Some of the military's famous surprise attacks were based on various deceptive practices meant to mislead the opponent and keep it calm, which then created a "surprise." The period referred to as "the cold war"[58] was riddled with such practices. Indeed, author Phillip Knightley finds it appropriate to quote Senator Hyram Johnson's 1917 proclamation that "The first casualty when war comes is truth" as the motto for his book.

For most people, truth means something nonproblematic, simple, straightforward, and objective; something which is "out there" and can be looked at, measured, described, perhaps photographed; an entity that manifests a genuine existence, regardless of human folly and manipulations. This perception of truth is a cultural belief that is also translated into special institutions dedicated, supposedly and in the popular perception, to the discovery and establishment of "truth," for example, courts and scientific research.

The type of "truth" that is constructed in courts, however, can be problematic. Courts are much more concerned with the application of a complicated set of politically constructed rules to create a version of reality that sometimes may be as far from the "truth" as is possible. Consequently, sometimes innocent people are convicted for doing things that they never did, or guilty ones are exonerated of crimes they actually committed. The language of "justice" and "fairness" is dominant in court, and it may sometimes contradict "truth." Nevertheless, one needs to be reminded that the type of truth or justice that takes place in courts depends on the quality of politically created rules on which court procedures rely. Likewise, the history of scientific research teaches us that "truth" can be a problematic issue there as well.

The digital age made this issue even more complex and thorny because it can make the production and reproduction of truth quite challenging. Digital technology enables the fabrication of reproductions representing different versions of reality not only on a mass scale but in a way that is difficult to authenticate.[59] And yet even with this technology that has created "versions" of reality at an unprecedented pace and volume, the belief in the existence of one truth out of many versions marks a solid foundation of our culture.[60] We are looking—for example—for genuine and authentic art and not fabricated products. We do not like to be told that a plagiarized product is a "genuine/authentic" one. We do not take kindly to people who take false credit. Thus while the modern age of digital and electronic technology may blur the difference between falsehoods and genuine artifacts or facts, we are still interested in the genuine, the authentic, and the truth. Without truth as a baseline, the existence of lies cannot be established. A culture that harbors falsified social realities must be wilder than Alice's wonderland. We will probably be lost there in a twilight zone where no clear boundaries between lies and truth exist, and there will merely be different versions of what reality may be like. This will be almost like a quantum reality. In this strange universe, reality may become a choice commodity. To us, it is inconceivable that anyone could live any meaningful or appreciable period of time in such a culture and remain sane (again, assuming that "sane" exists).

Prevalence of Deceit and Its Cultural Meaning

Claiming that deception is a cultural trait we all share still requires some attention to its prevalence. Indeed, Bailey (1991), Barnes (1994), and Robinson (1996) note that lies and deceit are ubiquitous and that individuals and governments are involved in deceptive schemes on a routine and regular basis.

Charting the history of public presentations of falsehoods, Robinson (1996) concludes that in the distant past public representations did not draw clear or sharp boundaries between truth and falsehood. Only in the modern era have such distinctions become important and significant and the contrast between true and false come into being.

British Museum assistant keeper[61] and author Mark Jones' volume (1990) shows how widespread different forms of deception (e.g., copying, transforming, embellishing, forging, faking) are in the arts,[62] as well as charting their causes, types, and modes of detection. Because it is difficult to estimate accurately the amount of deception, one is necessarily forced to examine its various manifestations. As pointed out earlier, the range of deceptions is impressively large and encompasses areas of human conduct from the personal to the collective, from pranks to military, and from the noncriminal to the criminal. Such a profound range must simply be taken as an indication of how prevalent deceptive behavior indeed is.

Peter Medawar, the cowinner of the 1960 Nobel Prize in physiology or medicine, produced in 1964 a few texts under the general title "Is the Scientific Paper a Fraud?"[63] Indeed, scientific papers do not reflect in an authentic way scientific practices. To get published, authors need to conform to accepted conventions. Thus one can argue that—like in other areas—researchers may manage to use some problematic practices as if they do not really represent some form of fraud.

DEFINING LIES: THE LITERATURE AND THE CONCLUSION

Having made some distinctions and discussed the issue of prevalence, as well as examining some cases, we are now in an advantageous position to clarify the analytical nature of the phenomenon we are looking at—lies and deceptions. As noted, defining "lying" is not an easy enterprise. Social scientist J. A. Barnes (1994) refers to a lie as "a statement intended to deceive a dupe about the state of the world, including intentions and attitudes of the liar" (11), disregarding the success or failure of the lie. Bailey (1991) contrasts "truth" with "untruth," pointing out that "untruth" has three basic forms: deceit, error, and fiction. He characterizes deceit as the "deliberate statement of what is known not to be the case" (127). Obviously, this definition relies heavily on "intention," which is itself another difficult concept because, among other reasons, honest mistakes and misunderstandings should not be included, not to mention the problem of measuring intent. Barnes suggests, "If we intend to deceive, we are acting untruthfully; if our untruthful act consists of making a statement intended to mislead, we are lying" (12), making the distinction between "lie" and "deceit" distinguishable by the intention involved and not the state of the world.[64] Vasek (1986) sharpens this approach further and notes that, indeed, "almost all writers include a definition of deception as the intentional communication of falsehood. . . . Deception occurs when the content of a message is untruthful" (272).

While truth may seem straightforward, in fact it is not always easy to establish. Barnes apparently enjoys quoting an 1897 statement he attributed to Mark Twain: "Truth is the most valuable thing we have. Let us economize it" (1994, 37). Bailey (1991, 128) notes that "truth" is always contextualized, negotiated, and dependent on power. The idea of "truth" being problematic, negotiated, and forever changing is typical of many contemporary, popular, and powerful paradigms such as postmodernism and relativism. Some of the works of such contemporary cultural icons as Michel Foucault, Hayden White, and Richard Rorty seem to blur the lines between "truth" and "untruth." Indeed, philosopher Christopher Norris's work (1996) criticizes—very strongly—such works for failing to provide clear demarcation lines between untruth and truth.

In view of these developments it becomes important to note that "the world" is not always composed of equally valid "versions," but that there are numerous forms of lying and deceptions and there are cases where one may find a "truth." Death is a good illustration. In fact, without "truth" such words as deception, hoax, concealment, lying, cheating, may become devoid of meaning. Truth and untruth span a wide range indeed, from simple acts like dyeing one's hair, using makeup, and resorting to plastic surgery to mask the signs of one's real age, through a physician's hiding from patients the true nature of their disease, to major military and political deceptions, as well as deception on the interpersonal level.

One of the better and more thoughtful approaches to lies and deceptions[65] was developed by Ekman (1992, 28). He states that a lie or deceit occurs when "one person intends to mislead another, doing so deliberately, without prior notification of this purpose, and without having been explicitly asked to do so by the target."

He then distinguishes between *concealing*—when a "liar withholds some information without actually saying anything untrue"—and *falsifying*—when the liar not only withholds information but "presents false information as if it were true" (1992, 28). While concealment may seem to be less disreputable because it may not involve inventing untrue accounts, it is quite capable of misleading audiences to believe in a false reality.

For example, while the late cellist Jacqueline du Pré has been portrayed as "the golden girl of classical music: beautiful, talented and irresistibly vivacious . . . known to her fans as 'St Jackie,'" that portrayal may have actually been mythical. According to a later report based on accounts provided by her brother and sister, Piers and Hilary du Pré, she was "spoilt, manipulative and emotionally dangerous to know . . . [her] family are tired of the image of her saintliness and want to set the record straight. . . . After Jacqueline's death her family came close to burning all her possessions in an attempt to 'exorcise' her influence." If indeed true, the concealment of du Pré's behavior was, at the very least, certainly helpful in crystallizing her image as a "saint." Indeed, du Pré's fans are reported as concerned that while no one challenges her musical talents, her saintly image may now be replaced by that of a monster.[66] Interestingly, what we have here are two diametrically opposed versions about du Pré; an outside observer could quite possibly produce a third version, which might lie somewhere between the two. In other words, the empirical basis behind the different constructions could provide more than two discrete black/white interpretations of du Pré.

Ekman bases his inclusive definition on a few elements. First, an *intention* to deceive must exist. This element is mentioned by almost everyone who has done work in this area, and it provides a good handle for distinguishing between benevolent and malicious lies. While "intention" is not always easy or simple to measure or establish, it appears that defining lies without it may be too difficult. Nevertheless, it needs to be pointed out at this stage that a possible alternative criterion to "intention" can be based on "outcome." This alternative is not problem-free either. One can indeed try to infer from behavior or performance what might have propelled this behavior into reality and thus attribute intentions. Such interpretations, however, can be open to criticisms and debates because behavior may not reveal direct intent and the thorny issue of "unintended consequences" may always lurk in the background. Second, this intention means that the deceiver or liar had a *choice* to make (whether to lie or not). Third, the target of a lie or a deception did not give his or her *consent* to be misled and the liar did not give any *prior notification* of the intent to make a false presentation. This is an important point because, as mentioned earlier, no one would consider calling film or theater actors "liars" even though they, by both choice and intent, make false presentations.

Ekman devotes an entire chapter in his book (1992) to examine the question of why lies fail.[67] The reasons he gives include bad lines (in the form of unpersuasive accounts), the difficulty in lying about feelings, fear of being caught, and guilt about deceiving. Of course, maintaining that lies fail accords well with our Western ideas about the immorality of lies. It is not at all clear, however, that lies "fail." Certainly, sometimes they do, but not always. The systematic and persuasive empirical evidence for deciding whether lies fail or succeed is still not available, and what we have most of the time are instructive but nonrepresentative anecdotes.[68] Nevertheless, and as so many authors have pointed out (including in this book), lies and deceptions form such an intimate part and parcel of cultures that it is reasonable to assume that very many lies probably do not fail. Besides, how are we to explain the fact that lies are so prevalent if they tend to fail so often? However we refer to them, situations in which there is a considerable disparity between factual reality and its presentation are probably extremely common. And this disparity is not just a random and innocent occurrence but the result of deliberate intent to deceive. In fact, we are willing to risk contradicting Ekman by suggesting that lies and deceptions fail far less than we would like to assume. In other words, "most liars can fool most people most of the time."[69]

Finally, the distinction between *concealment* and *falsification* is indeed illuminating. As Ekman points out, liars, if given a choice, will always choose concealment rather than falsification, because concealment is much easier to accomplish and liars tend to assume that it is less reprehensible than falsification (1992, 29). Ekman notes that there are many social situations in which we need to conceal, for example, poker or other and similar games, commercial and political negotiations, embezzlers who have to conceal the fact that they steal money, plagiarists who have to conceal the fact that the work they present as their own is really not. As is easily seen, these very same situations may give rise to falsifications as well. He is also quick to note that in some situations concealment is crucial for survival of those who opt or are constrained to lie, for example Jews hiding their identity under Nazi occupation, or spies. Such mass concealment of identity has been historically documented, for example the "Marranos" or "Anusim," Jews who pretended to convert to Christianity to avoid death during the Spanish Inquisition while secretly maintaining some aspects of their Jewish identity.[70]

Focusing on illustrations from areas such as religion, science, and politics, the issue of "people believing falsehoods" occupied much of Robinson's work (1996). Proceeding from creation myth to heroic legend (296), he states that both "are but two examples of major cultural institutions where truth was initially received truth" (299). Robinson takes the fact that people believe major falsehoods as a central trait of contemporary cultures. Indeed, these beliefs create difficulties in establishing clear lines between truth and falsehoods. Robinson's meticulous work (1996) is thus focused on the role of lies and deceit in representing realities and creating the illusion that a coherent culture in fact exists. His work obviously implies that lies and deception are main characteristics of cultures. Relying on Schwartz et al.'s 1988 dictionary, he maintains that to lie means "to utter a falsehood with an intention to deceive; to give a false impression" (1996, 26) and that "[t]he contrast between truth and falsity is the crucial distinction for the utility of the representational function of language" (27). He then distinguishes among five different approaches to defining lies. One measures the deviations from truth (characteristic of this book). The second focuses on a lingual analysis of words of deception. The third examines prototypical cultural lies, that is, meanings of words that "are based on prototypes rather than on distinctive features" (32–33). The fourth approach takes a wide-angle view to include all possible forms of deceit and concealment, stressing that the ability to choose (e.g., personal freedom) is essential to deception; that is, the liar must have the choice not to lie. The last approach is philosophical. It explores those acts that are intended "to make others believe what we ourselves do not believe" (34) and tries to understand the historical development of lying. Robinson himself is quick to warn us to beware of this approach because "[p]hilosophy is prone to develop a paralysis through analysis, and its practitioners are liable to forget the ultimate importance of synthesis, decision making and action" (38). If anything, his work needs to be taken as an indication of the complexity of this topic.

Barnes points out that the last place we would expect to find deceit is scientific research, and yet deception and misinforming subjects as to the real goals of the research have existed in science in at least two forms.[71] They can exist as deliberate acts in the research itself. For example, the research design by social psychologist Leon Festinger (and others)[72] on dissonance theory. Much more extreme forms can be found in some most famous research efforts, with far-reaching consequences, using designs that employed deception and misrepresentation. Some of these research efforts managed to place volunteers in difficult and compromising situations. Let us remember who some of them were. One of the first was the now classic study by social psychologist Solomon Asch (1955) who mixed innocent volunteers together with impostors that pretended to be volunteers but in fact were Asch's collaborators in a study that aimed to examine the effects of group pressures. Psychologist David Rosenhan (1973) required that people who were not mental patients be admitted to a mental hospital by pretending to be mentally ill. Rosenhan wanted to examine the process of what it takes to be admitted to a mental hospital and how the medical staff there reacted to and processed "patients" who were actually normal. Was consent from bona fide patients given? What about the medical staff being lied to and cheated? Psychologist Philip Zimbardo (1973)[73] placed innocent volunteers in a make-believe prison situation, where nonguards were required to play real guards. This experiment had to be abandoned after a few days instead of the two weeks it was planned due to brutalization and mistreatments of the volunteers.[74] Two research designs became perhaps more famous because of the extreme situations into which research subjects were put. One was social psychologist Stanley Milgram's work (1974), in what is considered now by many as a "no" research design. Milgram placed innocent volunteers in an experimental design in which they faced instructions to deliver electric shocks to other human subjects. In fact, no one received any shock, but the innocent volunteers were led to believe that they were in fact causing pain. The goal of this important research was to examine issues of obedience to authority. The other research was done without the subjects even being aware that they were part of a research project. This is the research done by sociologist Laud Humphreys (1972, 1975). Without getting any permissions, without making his targets of observations aware that they were being watched for research (in fact he misrepresented himself and lied to his subjects, and he used car registration numbers to track identities), Humphreys examined homosexual interactions in public restrooms and later tracked the men he watched and interviewed them. The other possibility is when the researcher deliberately deceives audiences by providing falsified reports. Such reports can range from reports on observations which have never been made, to manipulative, selective, and suggestive interpretations of data that are intended to deceive audiences into believing that a particular reality construction is the valid one.

Relying on systems theory, anthropologist and linguist Myrdene Anderson suggests that deception "[c]an be understood as simply the sending and/or receiving of misleading information in open cybernetic systems" (1986, 324). Consistent with Ekman's formulations, Anderson points out that there is an interesting interaction between deceit and secrecy. Thus deception can result from providing partial information, hiding crucial information, or telling factual information in such a way that no one will actually believe it (Anderson 1986, 325).

It is important to note that sometimes faked information can fall into an area where it is difficult to establish intent to deceive. For example, faked memories of imaginary traumas include many instances not only of untruths, but of tales that, despite their dubious validity, may receive full support from clinicians.[75]

Using "usefulness" as a utilitarian-defining criterion, Barnes (1994, 147) points out that lies "are a mixed blessing," and Bailey hastily adds that "some forms of untruth are admired" (1991, xvii). This may be so because the ability to lie and deceive may have some advantages in an evolutionary context[76] among humans, animals, and plants. Indeed, psychologists Robert W. Mitchell and Nicholas S. Thompson (1986) devote their edited book to examining deceptions among humans and nonhumans alike from that perspective.

Summarizing the different definitions and approaches, we would take lies and deceptions to form a bona fide reality when the following conditions are met: the existence of the intent to present a false front based on a choice to do so; an unsuspecting and uninformed audience; and the lies and deceptions are based on concealment or falsification. Once the existence of deception is established, we will need to distinguish between factual and interpretive truth (and lies).

Deviant Researchers: Truth, Deceit, and Fraud

Discussing fraud requires that we delve not only into some concepts about truth,[77] but also examine some cases that may fall between truth and deceit. Bailey (1991) states that the psychological, functional reason we "need" truth is "a universal psychological need to be sure of something, to put an end to the questioning, to know where one stands" (124). Put differently, the insecurity, powerlessness, and uncertainty that seem to plague modern life propel many of us to search for objective, perhaps eternal, truths, truths capable of establishing trust and integration. The quest for truth can be thought of as a quest for meaning (and religion as a sort of "ultimate truth" not to be questioned). Accepting Bailey's idea that on the personal level we are prone to fall prey to deceptive schemes of persuasion means that we may have answered the question of why we are willing to accept some of the most bizarre belief systems whose truth claims are certainly not empirically based. Many so-called religions and ideologies, totally based on pure belief, fall into this category. The ability of these old, or new, religious forms and ideologies to attract new recruits depends on the ability of recruiting agents to contextualize and tailor the specific solution they market as a remedy for uncertainty to specific innocent buyers. Much of their recruiting must rely on deception: telling only part of their "truth," hiding information, creating an atmosphere where suspension of natural disbelief occurs, and creating trust in the converting agents. For example, the typical price of specified salvation solutions in economic, social, and psychological terms is hardly, if ever, discussed. Regardless of this need, cultures are characterized by maintaining some level of lies and deception by relying on narratives such as mythologies.[78] In this respect, "truth" is perceived as the holy grail of scientific research.

It is beneficial at this point to distinguish between at least two types of truths. One type is *factual truth*. This is the truth involved in finding out whether something happened (or existed) or not. This type of truth typically does not involve interpretations, and if all the facts are known it is easy to establish. Questions such as "Who sank the *Lusitania* on May 7, 1915?" or "Do we inherit acquired traits?" or "Do N-Rays exist?" have factual answers. There is, however, a more elusive and difficult-to-establish type of truth, and that is *interpretive truth*. This type of truth may be based on combining facts in different ways. Political claims, ideologies, social identities, romantic involvements, these are all illustrative of this type of truth. Interpretive truth typically takes as building blocks some factual truths and integrates them into what looks like a coherent, systematic, and credible narrative. Issues regarding the selection of facts taken into account, their differential weights, and the way they are combined lie at the foundation of many mythologies, fairy tales, legends, fiction, and a few other forms of collective behavior. In fact, biographies or history books present interpretive truths, sometimes even when the research is careful and extensive. Early astronomers created an interpretive truth about their observations. These were often influenced by their worldview, but they were also influenced by the limits of their ability to see or measure. Why there were researchers who took factual truth and socially constructed meaning from them, by building "thick descriptions" ala anthropologist Clifford Geertz (1973) around these factual truths, is an intriguing topic. Many times it had to do with the motivation of those scientists to prove their worldview. Lysenko probably provides a good example, but then many would argue that he was not a scientist. Many Nazi scientists and physicians would qualify for this as well.[79]

Anthropologist Frederick George Bailey adds a fascinating dimension: "truth and deceit are inextricably entangled with power" (1991, xviii). For example, he points out that lies protect the weak from the strong and thus myth is actually developed by the weak as a weapon. An interesting question is whether a characteristic of research scientists who commit fraud is that they occupy weak positions. Author Jean-Francois Revel's work (1991), in which he claims that we live in an "age of lies," supports Bailey's claim that power plays a major part in deciding what is and what is not "truth," and Dreger's 2015 critical analysis of the liberal politicization of science adds more credit to this claim. For example, when Grote Reber—a radio engineer from Wheaton, Illinois—suggested in the early twentieth century that we could use the radio part of the spectrum to observe the universe, he was rejected and ridiculed by powerful figures in the academic astronomy community who were committed to optical astronomy. Truth, according to Bailey, can imply "authority, dominance, an unyielding absolutism" (3).[80] He states that the only way we

> can understand "truth" and "untruth" is to see them as rhetoric, as concepts used primarily for persuasion. They are political words, weapons for use in competition for power. In that context "truth" and "untruth" represent a tension, a tug of war, a dialogue between adversaries, who use these words in the contest, each striving to make their own ideas and values prevail. (128)

This incredible claim can be taken to imply that making a distinction between truth and untruth depends only on power and that "truth" and "deceit" are completely context-dependent.[81] But to sustain this assertion one must restrict oneself to interpretive truth and not to factual truth.

A Question of Belief

Issues of interpretive truth and power open the door to examine one of the most curious phenomena among researchers: belief systems that warp research. Indeed, like hounds locking on targets, we have cases of researchers whose belief led them to make some most interesting and innovative discoveries. There are other researchers, however, whose belief warped their integrity and led them into some dark alleys. While one may argue that such cases should perhaps be viewed as innocent mistakes, we beg to differ. Not following proper methodologies, ignoring evidence, and cooking data are all deliberately misleading acts. Hence we tend to view such cases as falling within the field of fraud. Nevertheless, we gladly admit that there is something rather odd, even embarrassing, about some of these cases, and examining this particular category is well worth the effort.

Deceit and truth: factual vs. interpretive truth. There are blatant lies, those whose existence can be established because they typically contrast with empirical facts, both on the personal and collective (sometimes national) levels. But what happens when the lie is not so evident or easy to detect and verify? When the lie is structured along what may appear as slight but are actually significant distortions? Does the motivation to falsify and deceive in such a case matter?

Cases of complicated deceptions are easily found in military history, politics,[82] and infidelity. But what if a person or persons believe that the nature of reality has a certain form, and they discard or twist evidence to the contrary out of a belief in their own fabricated reality? In "their truth," shall we say? What if they fail to see or understand anything outside their "truth," for example, homophobes who—contrary to evidence—continue to believe that homosexuality can be "cured"? How about parents telling their children "white lies" because they feel that facing the truth is beyond their, or their child's, power and ability to process? Remember there are physicians who avoid telling some patients the true nature of their disease for similar reasons?

There are numerous illustrations of situations where people had such a strong belief in something that they were willing to discard, or deliberately ignore, contradictory (or not supportive) evidence. The area of military intelligence provides three examples. One is the American failure to discover in time the impending Japanese attack on Pearl Harbor (December 7, 1941),[83] another is the failure of Israeli intelligence to expose the Egyptian-Syrian attack of October 3, 1973 (the Yom Kippur War), and the third is the failure of the Allies to anticipate the German 1944 counterattack in the Ardennes (Battle of the Bulge). In all three cases the attackers inflicted temporary military catastrophes. Many observers point out that in these cases intelligence was a "prisoner" of its own "concept." While this "concept" assumed that the attackers had hostile intentions, it did not include an imminent war. Thus despite having enough information that a military strike was indeed imminent, military leaders ignored the information. In other words, Japan, Syria and Egypt, and the Nazi Wehrmacht were highly successful in deceiving their opponents because these opponents had faith in some given belief about what reality was like.

As we saw earlier in the section that discussed deviant knowledge claims, science, too, has its share. While physicists had enough evidence that it was possible to view the universe via radio waves, this information was discarded as unreliable in favor of an optical view, and radio astronomy was not allowed to develop.[84] The Ptolemaic[85] view of the solar system was accepted as a dogmatic (even sacred) truth despite accumulating evidence to the contrary. Earlier than that was the ancient belief that the earth was flat.[86] This ancient belief is particularly interesting because believers at that time did not suffer from a light pollution effect that so many people living in urban areas experience. When they came out at night they could easily see the curvature of the horizon. And yet despite this direct observation, many still believed in a flat earth. The acceptance of quantum mechanics and plate tectonics suffered a similar fate. As late as the 1980s, accepting the fact that an infection by the bacterium *Helicobacter pylori* was a major cause for peptic ulcers was not easy.[87] In fact, Australian physician Dr. Barry Marshall had to infect himself in order to prove this fact (and share the 2005 Nobel Prize in physiology or medicine with his collaborator Robin Warren). Each scientific discipline constitutes a belief system; within this belief system puzzles are raised, anomalies are ignored or denied, and the plausibility of various types of explanations and knowledge claims is weighed. The recognition of anomalies, acceptance of new ideas and the rejection of others, and the problem of "resistance to innovation" in science, are all directly linked both to science as a belief system and to the fact that researchers operate on the basis of, and within, an extensive and complex matrix of existing paradigmatic beliefs. This matrix includes the prevailing disciplinary professional beliefs and criteria for evaluating knowledge claims.[88] It is thus possible that a belief in false knowledge claims may take hold if these false claims somehow fit a worldview. Such cases probably involved Lysenko, Cyril Burt (suspected of falsifying his work in educational psychology[89]) or Caloric. We picked some instructive illustrative cases to follow.

Case 1: Mark Spector

This is the previously described case involving Mark Spector (McKean 1981[90]), the graduate student at Cornell University who "discovered" a revolutionary explanation for cancer that turned out to be a fraud. McKean stated that the fraud was possible because "scientists were eager to accept a theory that seemed to show a chemical chain leading to cancer" (19). Professor Racker and others were entrapped by their desire to believe in the type of theory that Spector's work developed. In an interview with McKean, however, Professor Togasaki, a former instructor of Spector's, noted that "[t]he whole (scientific) system is based to a very high degree on trust" and expressed his concern that scientists may become overly suspicious of each other's work (McKean 1981, 23).

Case 2: William Summerlin

This case was also described earlier and involved the famous "painted mice affair" in which William Summerlin, a researcher at Sloan-Kettering Institute, painted the skin of a mouse with simple ink and claimed that he had successfully completed skin grafts between genetically different animals. Summerlin was clearly determined to "prove" his claim, but he chose an odd "methodology" to substantiate his belief. What makes this even stranger is that Summerlin admitted he had inked the back of a mouse (Zuckerman 1977, 132), an admission that clearly paints this case with the unpleasant color of fraud. Reminiscent of the case involving Bjorn Lomborg (discussed elsewhere in this book), while a scientific investigation into this case found Summerlin guilty, he was convicted of misrepresentation and not of fraud (Culliton 1974). The question created by making such a distinction between these two concepts remains open, and its solution undoubtedly requires some innovative linguistic and cognitive exercises.

If Summerlin used ink, it is interesting to note that in 2006 Lingjie Zhao—a former doctoral student at the University of Iowa—was found to have falsified research by, among other deviant activities, using a marking device to darken lines in an auto-radiographic film. The digital age makes cheating much easier, and it may increase the prevalence of deceit.

Case 3: Paul Kammerer

This case, which is similar to the painted mice affair, involves the now famous "midwife toad" (Zuckerman 1977; Koestler 1972; Asimov 1979). Paul Kammerer was an Austrian biologist who strongly believed in the Lamarckian view that acquired characteristics could be inherited. From 1918 he worked with toads and salamanders in an attempt to validate the theory. Males in some species of toads possess darkly colored thumb pads.

The midwife toad, however, does not have this characteristic. Kammerer manipulated environmental conditions in a fashion that he hoped would create a midwife toad with darkly colored thumb pads. Since this did not occur, Kammerer painted the thumb pads of his toads with India ink. Kammerer was very secretive about his experiments, and detection of fabrication was not easy. It was finally discovered, however, and Kammerer was discredited. After the exposure Kammerer committed suicide.

Case 4: Ferdinand Moissan

This case concerns the French chemist Ferdinand Frederick Moissan,[91] winner of the 1906 Nobel Prize in chemistry, who undertook in 1893 to show that he could create artificial diamonds (Asimov 1979, 176–77). Today we know in detail the conditions under which graphite will change into carbon and how diamonds can be created. This was not known in 1893, however, and Moissan, with the technology and knowledge available to him at that time, could not possibly have created diamonds. The only problem is that his experiments did result in the creation of diamonds. Small, impure, and colorless, but diamonds nevertheless. Historians of science generally assume that one or more of Moissan's assistants wanted to help him, or perhaps to play a joke, and introduced the diamonds into the experimental set. Moissan fell for it and the assistant probably could not back out.[92]

Case 5: Prosper Blondlot, Lysenko, and Kinsey

The fifth case concerns the previously mentioned "discovery" of N-rays by Prosper Blondlot.[93] This case is particularly interesting because one could argue that the case of Blondlot perhaps reflects a perceptual and cognitive bias resulting from wanting to see something that is not really there. While there is some merit to this view, in essence we believe that science is not based on wishful thinking but on paying careful attention to equipment and observations. We may want to forgive Blondlot for what may have been his burning desire to find some new rays, but that forgiveness does not discount the fact that he built an entire contraption that measured nothing and reported that this nothingness "produced" something he referred to as N-rays. Jules Verne might have had fun with this, but we are observing science and not science fiction. Luckily, the damage caused by N-rays was nil.

Another related and famous case is that of Kinsey. That his work had a significant impact is not debatable, but was that impact beneficial? And what is one supposed to do with Lysenko, who—courtesy of the Soviet regime—invented a nonexistent genetic science and caused real damage to Soviet genetic research and very serious misery to anyone who objected to this "science"? Did Lysenko—contrary to evidence—really believe in the inheritance of acquired traits? The Lysenko case is not unique. A similar case involves creationists who believe that the Bible accurately and completely describes the creation of life on earth and have devoted a great deal of energy to refuting and denying evolutionary theory. Perhaps the worst case of research in this line was the existence of Heinrich Himmler's cultural research unit Ahnenerbe. Researchers in this unit set out to use archaeology, anthropology, and cultural studies to establish the existence of the Aryan race. And they did believe in what they were doing, including the atrocities some of them committed (see, e.g., Pringle 2006). We tend not to include cases involving honest mistakes, but cases that involve personal or ideological biases of researchers who wanted to prove some belief and forgo proper methodologies to do it cannot be included in this category

Case 6: Cyril Burt

This is the case of a psychologist, Cyril Lodowic Burt, who taught at both Oxford and Cambridge Universities. Burt studied the occupational status of parents and compared it with the IQs of their children. He found a percent match between IQ grades and occupations. The higher the parent ranked in the social stratification system, the higher the IQ of their children was likely to be. Professor Burt died before his studies were investigated.

Suspicion about Burt's work was aroused as early as 1972 when Leon Kamin, a social psychologist, read Burt's studies and immediately sensed that they were fraudulent. It was not until the September 29, 1978, issue of *Science*, however, that psychologist D. Dorfman proved beyond a reasonable doubt that Professor Burt fabricated data (Weinstein 1978; Asimov 1979). Asimov (1979, 139) noted that in this case as well, Professor Burt probably falsified data so that they would fit his theories. Author Nicholas Wade (1976, 918) commented that "Kamin's interpretation was that Burt's data remained unchallenged because they confirmed what everyone wanted to believe."[94] The 1990s saw a revival of this case. Sociologist Ronald Fletcher published *Science, Ideology, and the Media: The Cyril Burt Scandal* (1991), in which he questioned and criticized Burt's critics and presented a strong defense of Burt's work. Then in 1995 N. J. Mackintosh edited *Cyril Burt: Fraud or Framed?* which continued Fletcher's line and suggested that while his defenders were not always right, not all criticism of Burt's work was valid or justified.

Case 7: Jay Levy

This case involves Jay Levy, a research assistant of J. B. Rhine, a famous and respected ESP researcher. Levy was found tampering with the equipment that recorded the movements of experimental rats in order to make it appear that the rats had modified the operation of a random number generator, thus proving the experimental hypothesis. Apparently, Levy wanted Rhine's work in ESP to succeed at all costs (Zuckerman 1977, 116).

Reporting Experiments

This case involves an intriguing research report by social psychologist Leonard Berkowitz (1971). Berkowitz looked up how seventeen social psychology textbooks reported a specific experiment by Stanley Shachter on deviation and rejection.[95] He found that few texts had given an accurate report of the experiment. Most of the textbooks reported only portions of the conclusions, thus distorting the experiment's original meaning.[96] It appears that the authors simply selected those aspects of the experiment that supported and fitted the textbook's general theme and disregarded such details as accuracy, validity, and reliability.[97]

Berkowitz's work is more than forty years old, but Joseph Simmons, Leif Nelson, and Uri Simonsohn's more recent work (2011, 1359) points out that "flexibility in data collection, analysis, and reporting dramatically increases actual false-positive rates. In many cases, a researcher is more likely to falsely find evidence that an effect exists than to correctly find evidence that it does not."

Many of the cases brought so far were chosen because they illustrated a clear cognitive intention to commit a deviant act of fraud: to fabricate or invent data or use fraudulent research and methodologies. Obviously, researchers can make honest mistakes with no cognitive intention to cheat[98] and even—as Hamilton (1996) points out—believe in some shady claims. The area, however, between intentional fraud and simple honest mistakes may sometimes be gray and may include cases that are difficult to classify. For example, in 1974, John Gribbin, a doctor of astrophysics, and Stephen Flagermann published the book *The Jupiter Effect*. In it, Gribbin put forward the idea that an alignment of the planets in 1982 would lead to increased solar activity and cause great earthquakes on our planet. Los Angeles was expected to be the site of a massive and devastating quake. Gribbin was immediately attacked by various astronomers, who remarked that the book was "astrology in disguise" and "pure fantasy" (Gardner 1981, 345). In the July-August 1979 issue of *Mercury*, Belgian astronomer Jean Meeus debunked the so-called "Jupiter effect." Lipton published a paper in January 1975 in the *Griffith Observer* titled "The Great Earthquake Hoax" denying the theory's validity. Eventually, in the June 1980 issue of *Omni*, Gribbin publicly retracted the idea, admitting that the whole basis of the 1982 prediction had been proven wrong.[99] Was all this an intentional hoax (as Lipton feels) or an honest scientific mistake?

Indeed, Shapin's (and others) works on the history and sociology of science illustrate how science is not free from various cultural and political influences.[100]

Senior scientists in charge of junior scholars are often vested with two major responsibilities. The first is to be the actual gatekeeper of science, to monitor and closely examine various types of innovations.[101] Many of them believe in the accepted and established scientific paradigms and spend much of their life proving them or working them out. Many of these researchers could be expected to have a natural propensity to suspect new ideas and to encourage ideas, experiments, and findings that reinforce the paradigm they believe in. They can thus bias younger scientists toward theories and findings that support their scientific worldviews and to reject, at least initially, new or contradictory information. Two relatively simple illustrations are Moissan and his artificial diamonds and Jay Levi and ESP. A bit more complicated case is, perhaps, the case of graduate student Mark Spector, where "'scientists were eager to accept a theory that seemed to show a chemical chain leading to cancer.' Prof. Racker and others were entrapped by their desire to believe in the type of theory that Spector's work developed." (McKean 1981, 23).

Deception as a generic quality and fraudulent research: summary notes for this chapter. As noted earlier, what makes us able to be involved in complicated deceptive schemes is our human ability to use a highly symbolic form of communication to transcend the limitations of time and place. Although in itself this is not a good reason to utilize these abilities, most researchers of lies and deception are convinced that all of us have been involved in various forms of lying and deceit. The reason is that the cultures we have constructed have lies and deceptions built into them. Processes of socialization involve acquiring the ability, reasons, and practices of lying and deceiving. The boundaries between "deceiving" and "not deceiving" are not set as absolute qualities. Rather, we learn more about such issues as how much to deceive, in which situations it is appropriate to utilize what type of deceptions or to whom one is allowed to lie. Lying is thus not a matter of principle, but of degree and quality. The answer to the question of "why people lie" can be found in its opposite. That is, people lie because they can and because we live in cultures that incorporate and normalize the practice of lying and deceiving into everyday life as a required and accepted behavior. Lying and deceiving are therefore both a personal ability and a salient cultural structural characteristic. Deceptions thus emerge normally when structural opportunities or pressures in the social structure of cultures deem it necessary and make it possible. Perhaps the more interesting and proper questions to study are why and when people do not lie.

Given this ability and practice, deceit has emerged as a major characterization of human cultures, substantively and historically. Bailey (1991) and Barnes (1994) provide anthropological evidence for cultures that are characterized by high levels of deception and lies and what these practices led to. Living with false ideologies is, however, nothing new. Indeed, historian Perez Zagorin (1996) offers further evidence for the historical significance of lying and dissimulation. As Robinson (1996) argued, many cultures did, and still do, live with central untruths and their members believe, for example, in deceitful mythologies.

While the distinguishing characteristics of deception are the *intent* and *choice* to mislead *unsuspecting* audiences, deception itself can assume many and varied forms. To help complicate things a bit further, we find in many instances a reluctance to designate a deceiving behavior as "deception," preferring instead to use other descriptive words so that the *intent* to deceive becomes cloaked and obfuscated. Indeed, psychologist Paul Ekman's work (1992, 28) points out that the terminology we use is both important and controversial. For example, not everyone would want to use the word "lying" in the context of "concealing"; some—he says—reserve the word "lying" for "bolder acts of falsification."

Most of us, most of the time, would not wish to be associated with a motivation to deceive, which we tend to view with scorn. This intent, after all, can be easily interpreted as a motivation to cheat and lie, which taints behavior unpleasantly. Nevertheless, it would probably be impossible to maintain social life as we know it without deceptions and lies.[102] Likewise, nonhuman deception (animal and plant) is crucial for the survival and thriving of some species. Being so prevalent, why should we then object to the imputation of deception? Plainly, because an incommensurability (Cohen and Ben-Ari 1991) is created here. On the one hand, practicing deceptions can be important and useful. On the other hand, we praise "truth" as an important value. It is this contradiction that fuels our ambivalent attitudes toward deception, including our hesitation to use the term more boldly. Moreover, deception in research, or using a scientific cloak to mislead, creates a particular incommensurability. The perception of what scientific research is all about—that is, some truth—is not a conducive context for deceptions.

Deception in scientific research presents, then, a different perspective. While deviant knowledge claims may be one potential door to introduce changes into science, deviant researchers—that is, those that falsify, fabricate, and deceive research—present the ultimate contradiction and violation of the ethos of scientific research. Such deviant practices violate two crucially important facets of science: trust and loyalty, and they consequently cause a suspicion of betrayal to emerge. Eventually, scientific research that is based on deceptive practices will disintegrate as its relation to reality, and its ability to solve riddles, will—sooner or later—crumble. Nevertheless, the price paid for such practice may be much too high.

Bailey (1991) points out that:

> Untruth—the whole range from outright lies to fantasy—is the mover . . . when we want to move the world . . . we need deceit and fictions . . . creativity, imagination, fantasy, ingenuity, charisma—all those denials of the iron cage . . . in rational bureaucracy and which are constitutive of any basic lie. (126)

He further notes that:

> [G]rand designs like capitalism or communism or Islam or Christianity, or lesser designs like functionalism or structuralism or postmodernism, and other such basic lies, might all in the end be subject to the same kind of empirical test. . . . The ultimate control on a design (that is, a purported truth) is whether or not it copes with a real world of fundamental problems. . . . Truth, perhaps, is what works. (122–23)

Without explaining why, Bailey conveys his conviction that in the long run, a basic lie stands a good chance of being exposed even though it may often be presented as a basic truth.

While the relevant literature presents a decent number of detailed descriptions of fraud in research, the main innovation in our work lies in the perspective and the context. Our major professional aim has been to examine phenomena of deviance in a way that contrasts them with the nondeviant within the relevant organizational context. Such presentation can help us understand not only both, but also the nature and location of the moral and organizational boundaries that demarcate the "right" from the "wrong," the "genuine" from the "sham." This in turn helps us develop a powerful analysis, and hence better understanding, of the role of deviance and nondeviance in key processes of social and organizational change and stability. In the context of this book, we examine how misrepresentations of scientific data can be understood within the organizational structures within which they take place. Using the term "fraud" in the context of science thus does not mean only some technical intentional deceit. As law professor Stuart Green (2006) points out, it also carries a moral meaning, that of dishonesty and dishonor.

Many possibilities exist for approaching the subject of truth and deception, just as there are many manifestations of truth and deception. In this so-called postmodern age, some have even replaced the rhetoric of "truth" and "deception" with that of "versions." This last approach tends to problematize the existence of truth and falsehood and has a tendency to present different "versions" of reality; the validity of each is sometimes judged to be more or less the same (excluding, sometimes, the particular and idiosyncratic version of the specific claim maker, of course). This is not the approach taken here. Using the language of truth and falsity is simply essential for discussing issues of fraud in research. The line dividing truth from that which is not truth can, and should, be drawn. The meaning of drawing that line, as well as the ability to draw it, are issues that relate both to power in the organizational context and to morality. In the empirical cases presented and discussed in this book, this approach should be obvious.

One general and important observation arrived at this point is that we have to understand that the personal morality and beliefs of research scientists can, and sometimes does, affect not only the choice of their studies, but the results as well. That such factors can influence the choice of a research project is less problematic. But when ideology affects the findings or interpretations, it is a much more serious problem. Such famous persons as Sigmund Freud (Masson 1984) and Carl Jung (Noll 1994, 1997) were involved in exactly such practices.

Kinsey, Blondlot, Lysenko, and a host of other research scientists not only allowed their personal beliefs to affect and lead their research, but also used fraudulent practices. For example, Kinsey selected his samples in such a way that the evidence he created and presented in public gave a "scientific" picture of human sexuality that may have been totally warped and skewed. Practicing such tactics deceives unsuspecting audiences and aims to redefine the boundaries of their respective "imagined communities." While Kinsey desperately wanted to change those boundaries, however, other scientists may have wanted to keep those boundaries intact. Some of these attempts (not necessarily by selecting subjects), one must add, were quite successful. Lysenko, for example, chose to brutally and actively ignore and suppress contradictory information (as well as its bearers).

The collective data analysis in a previous chapter and the variety of cases presented in this chapter illustrate, in no small terms, that when the relevant information is available, a clear line between "truth" and "deception" can be drawn. This is probably valid as well for other areas of life where deception may occur. Another important conclusion is that—contrary to honest mistakes—the ***intention*** to falsify, lie, and deceive an unsuspecting audience is a crucially distinguishing hallmark of deception. Deliberately ignoring normative and accepted methodologies may sometimes also be taken as an indication (e.g., Blondlot, Kinsey). Nevertheless, and just to illustrate how complex this issue can become, we introduced the case of Blondlot. Did he *intentionally* fabricate his apparatus? On one level, yes, he most certainly did. The lenses he put in his contraption were aimed to capture something that did not exist and was not there. And, indeed, they did not capture the elusive and nonexistent N-rays. That fact should have been clear to him from the day he put together the apparatus that was supposed to "trap" the mysterious N-rays. On another level, one can argue that it appears that Blondlot may have genuinely believed that his contraption did capture those rays and therefore may have lacked the intention to falsify evidence. The problem, of course, may lie with our difficulties in establishing "intent." Moreover, the case of Blondlot illustrates another difficulty so common in social construction theories.[103] Blondlot was working in a paradigmatic context that was conducive to the development and belief in N-rays. As the famous scholar Bertrand Russell has noted, the application of a mathematicized astronomy leads to a mathematicized universe. Likewise, optical astronomy created severe obstacles for the acceptance of radio astronomy as a bona fide science.[104] In other words, what we see, perceive, and understand is mediated by a few filters, one of which is culture, or a paradigm in the case of scientific research. This is another illustration, if you like, of Thomas's now classical dictum that if situations are defined as real, they are real in their consequences. But the "solution" to this problem lies in the N-rays story itself. When a hard, cold, empirical test was applied to Blondlot's "experiment," the reality that was based on falsehood simply evaporated into nothingness, regardless of what Blondlot believed. So ultimately, even the case of Blondlot can be used as an illustration of how a reality based on falsehoods can be collapsed.

FIVE | Deviance in Scientific Research: Norms, Hot Spots, Control

This chapter examines the specific form of fraud in science in the general context of deviance and the organizations in which this form of deviance takes place. We begin by examining the issue of fraud in research and some organizational responses, and we continue to critically look at Merton's scientific norms. We then move on to discuss social control in scientific research. Once we have clarified what the normative and control aspects are, our next section examines potential forces that are conducive in helping to make a researcher deviant. Once a researcher crosses the line and gets involved in deviant practices such as fraud, falsification, fabrication, then the question becomes how do such persons account for this? Consequently, our next section looks into illustrations and types of accounts, followed by a section that provides illustrations of famous researchers who practiced or committed fraud. Given all these examinations and analyses, we develop a discussion that places the specific form of deviance—fraud in research—within more general theories of deviance. We argue that practicing or committing fraud in research does not merely reflect a sudden and unexpected move. It reflects a process of becoming. This process means that a nondeviant researcher has to pass through a series of cognitive steps the culmination of which is that this scientist becomes a deviant. This argument, almost naturally, leads us to the next section, where we ponder how widespread is the phenomenon of fraud in research. Given the data we collected and our analysis, we are clearly inclined to argue that it is probably closer to a tip-of-the- iceberg phenomenon rather than to a bad apple occurrence. The next two sections form a natural continuation of the arguments we developed to this point. The first focuses on the comparative issue of fraud in research and similar forms of deviance, and the next touches the issue of academic politics and fraud. We end this integrative chapter by examining one of our initial deliberations, which is looking at reactions to fraud in research as an issue of rhetoric and boundaries.

BECOMING DEVIANT AND ORGANIZATIONAL CONSIDERATIONS

It is always tempting to search for personal motivations behind various acts of scientific fraud committed by researchers. And, indeed, in the final analysis it is the individual researcher who decides to go through the gate and along the road that leads to falsifying and fabricating data and thus move from being a reputable nondeviant to a disresputable deviant scholar. Having said that, though, we are still left with the puzzling question of whether there are social or organizational patterns associated with such decisions. While we shall suggest a possible approach to understanding what may happen to an individual researcher as he or she moves through the door of deception, we also need to examine the organizational structure associated with such actions.

Indicating how, where, and when different research scientists have falsified data and reported nonexistent observations (and perhaps implying how such practices may be diminished) becomes a central issue. By dealing with such cases the research community can indicate to its own members and to the public that it has the mechanisms to detect and punish deviants and that it is a cohesive community with clear-cut boundaries that knows its moral-ethical obligations. Therefore it is in the interest of the research community to detect deviants, but not too many of them. Research is supposed to seek the truth. Too many liars, cheats, or deviants, and the very legitimacy of the scientific research endeavor can crumble. Nothing could be more dangerous for the scientific community than the discovery that a high percentage of its research members were engaged in fraudulent practices.

The tension between these two poles—feeling uncomfortable about reporting too many fraudulent cases but at the same time wanting to show that such cases are under control—can yield what may appear to be some strange results.

One illustration is the case of Bjorn Lomborg,[1] a Danish academic and scholar who won world renown for his 2001 book *The Skeptical Environmentalist*. Following the publication of this book a few environmental researchers filed complaints against Lomborg with the Danish Committees on Scientific Dishonesty (DCSD), accusing him of scientific dishonesty. On January 6, 2003, the DCSD reached an odd decision. It stated that *The Skeptical Environmentalist* presented:

1. Fabrication of data;
2. Selective discarding of unwanted results (resulting in selective citation as well);

3. Deliberately misleading use of statistical methods;
4. Distorted interpretation of conclusions;
5. Plagiarism;
6. Deliberate misinterpretation of others' results.

Consequently, the DCSD made the unavoidable decision that the book was scientifically dishonest. Nevertheless, it also decided that the person who wrote the book and whose name appears on the book as the author, Lomborg, was not guilty. Why? Because of his lack of expertise in the fields in question. One needs to read what looks like an Orwellian newspeak cop out and just keep the silence out of sheer amazement. Facing danger, an opossum will "play dead"; that is, it will ignore reality and danger. When the DCSD faced the danger of stating clearly without "ifs and buts" that a major falsification was committed (and risking the potential wrath of admirers of the suspect legal proceedings and other potential unpleasant exposures in social media), it played opossum.[2] That is "dead" morally.[3]

Unfortunately, while the field of deviant researchers is a most fascinating subject for scientific inquiry, it is also one that has been inadequately researched by relevant professionals in the areas of deviance or criminology. The fact is that fraud in scientific research is—primarily—fraud; that is, immoral, unethical, and illegal behavior.[4] Yet hardly any criminal investigations have been carried out in this area. Only recently, and hesitantly, have some cases been investigated. The lack of relevant developed theoretical and empirical knowledge in this field probably reflects both the long-standing reluctance of many researchers to face the criminal implications of this issue and the difficulties of investigating the subject. Processes that led to the establishment of the U.S. Office of Research Integrity (ORI)[5] were initiated only in the late 1980s. Thus only in recent years have researchers begun to hesitantly probe this mysterious and troubling terrain, and they have revealed some intriguing results. For example, replicating research results is problematic in some disciplines, and in too many cases replications fails.[6] Further, in many cases it is difficult or impossible to get original data from scientists, and many journal articles are retracted[7] as a result of fraudulent research. One must also note the strange fact that a community that is dedicated to the systematic recording of data has neglected for so many years to accumulate systematic and full data on deviant scientists and fraudulent research. Many of the cases cited in ORI's website are incomplete, lacking information such as how the deviant researcher responded to the allegations, what happened after the case, and how the relevant organization or institution responded.

Next it would be beneficial to explore some aspects of science and its structure with an eye to potential hot spots, situations where an increased probability of fraud exists.

Scientific Norms

Science is dedicated to the accumulation of "certified knowledge." So the question arises: Under what conditions and norms should the pursuit of "certified knowledge" be carried out? Sociologist Robert Merton (1973, 12–14) outlined the following four idealized norms,[8] which to him seemed to be central to the normative practice of science: universalism, that the validity and truth of scientific statements be totally separated from the personal characteristics of the person who makes them; communality, that scientific findings should be freely shared with others (secret or classified research is antithetical to the spirit of science); disinterestedness, that the researcher be guided not by personal motives (e.g., profit) but by the wish to extend scientific knowledge; and organized skepticism, that researchers be encouraged to openly, honestly, and publicly examine each other's work critically.

These four norms are energized (Storer 1977) by the need of researchers for professional recognition and by the interactions among networks of scientists that make up the cultural matrix of scientific research. Conforming to the four Mertonian norms should prevent deviance in science or fraud in research and guarantee that the scientific effort be oriented toward accumulating certified knowledge. So which factors and forces within science enhance conformity to the rules and which create pressures toward rule breaking and deviance?

Social Control in Scientific Research

Social control—formal and informal—in scientific research has two meanings. As we have seen in a previous chapter, one is the social control that is enforced by the organizations and institutions that practice science. These organizations and institutions create sets of ethical rules concerning the limits of normative behavior when doing scientific research, and they appoint social-control agents to make sure these ethical rules are adhered to. Such adherence can be encouraged by a number of activities, such as seminars and other educational programs, exercises, informal interactions among academics, the work of investigative committees, and the application of sanctions. We have already discussed this type of social control at some length.

The second meaning of social control can be found in sociological constructions of deviance and theories of criminology that focus on the individual.[9] Deviance, according to these theories, is not *caused* but is *made possible* because societal agencies or personal self-restraints can no longer prevent its occurrence. Deviance, according to this approach, is a natural state. All these theories, in one way or another, follow a basic Durkheimian approach that states that deviance will happen whenever the hold of societies and groups on the individuals becomes neutralized, broken, or weakened (Frazier 1976). The problem then becomes one of mapping the various control mechanisms and showing how they function and how they are rendered ineffective.

Theories of control in this context focus on two main mechanisms, which can function either separately or together. The first is an inner mechanism that can assume many names or forms, but that is essentially created through the internalization of norms and various other socialization processes. This mechanism can be labeled "super ego," "conscience," "self-control," or "morality." On the other hand, we have so-called external mechanisms in the forms of family, community, and control agents (i.e., police officers, judges, psychiatrists, or teachers). In principle, every control theory plays with these two mechanisms.

Social control in scientific research is no exception. Most researchers would agree that through the long process of training, research scientists internalize the norms of scientific inquiry and methodology. These researchers are socialized to and taught what types of questions to ask and what types of methodologies to use, and above all they are taught that their only guide should be the search for scientific truth. Cheating and fraud become the most serious infractions conceivable in the context of doing scientific research.

Sociologist Harriet Zuckerman (1977, 90) specifically states that "social control in science depends partly on scientists' internalizing moral and cognitive norms in the course of their professional socialization and partly on social mechanisms for the detection of deviant behavior and the exercise of sanctions when it is detected." What, then, are the external forms of control in science? As sociologist Deena Weinstein rightly points out, "[i]n science, as in polity, policing requires suspicion that a 'crime' has been committed, collection of evidence, and finally punishment of the offender" (1979, 647).

Since a police or a fraud-detecting unit does not exist in science, the first and most obvious mechanism to prevent fraud in scientific research is the norm of "communality" mentioned by Merton. Adhering to this norm means that researchers' actions, experiments, and results are visible. Many forms of deviance become possible because they are not committed in public. To commit successful fraud in scientific research requires that it be concealed at such a level (or form) of sophistication that even when fraudulent findings are presented publicly the probability that the fraud will be detected is minimal. The second mechanism is Merton's norm of "organized skepticism." This norm implies that research scientists are, a priori, a suspicious breed, trained to criticize each other and to search actively for defects, improbable statements, and bad methodologies.

The first and second mechanisms receive direct expression in the way researchers present the results of their research, whether at conventions and meetings or in published articles, papers, or books. The work is submitted for an open and thorough peer review, and its scientific merit is judged. Reputable professional journals would not publish works that have not satisfactorily passed peer review. Most journals solicit the judgment of two to four anonymous referees.

The third, and probably most important, external social control mechanism is replication. This mechanism stems directly from Merton's norm of universalism and is grounded in the theory that if experiment x yielded result y in one specific time and place, the same experiment should yield the same result in a different time and place. This is an obvious expectation in natural science research. Social science research presents a more complex picture. Some universal rules have been established in social science research, for example, positive correlations between frustration and aggression or between anxiety and performance (up to a point). Some research in social science, however, is context- and time-dependent (such as ethnographies). Given this dependency, outcomes of research may present a decent variance over time or place.

In the following section, we shall point out that social control in scientific research is much weaker than what people may think and that there are several structural, as well as personal, incentives in science to commit fraud.[10]

FORCES CONDUCIVE TO THE CREATION OF DEVIANT SCIENTISTS

The Mertonian Norms Re-examined

The first of these norms is universalism, meaning that the truth of a scientific statement is separate from the personal characteristics of the individual scientist. This ideal, however, may not always be possible. The very process of discovery, of invention, of innovation, or of developing a new theory or interpretation is clearly dependent on the personal, and especially on the psychic, makeup of the research scientist involved. Philosophical theories, conceptualizations of social orders and changes, are often linked to the worldviews of the particular scientist. A Marxist sociologist and a Parsonian sociologist would not explain the Russian Revolution similarly. Moreover, interpretive paradigms place a serious strain on this norm. Two of the buzzwords used to evaluate scientific interpretations in the social sciences, for example, are whether a particular explanation is "persuasive" or "interesting." But what makes something "persuasive" (or "absurd" or "not serious") or "interesting" is not entirely clear. Some feel that postcolonial interpretations are more persuasive than a functional interpretation, but are they really? What may be accepted as a bona fide and persuasive interpretation is further complicated by the fact that words are a central tool used in all interpretations. For example, choosing more dramatic words meant to emphasize a point but not to falsify the findings may create an impression of exaggeration, but is it misconduct?

The second norm, communality, is easy to challenge as well. According to this norm, researchers should freely share their findings with each other, a norm that is violated almost daily for a variety of reasons. The most obvious example is research that has been declared secret or classified by a government. In other cases, researchers in fiercely competitive fields fear that others will steal their ideas, and they become protective of their theories and findings. Despite the norm of communality, gaining access to original data is extremely difficult. Deena Weinstein (1979) and author Royce Rensberger (1977), for example, report on an Iowa State researcher who requested the raw data from thirty-seven authors whose works were published in psychology journals. He received replies from only thirty-two researchers, of whom twenty-one claimed that their data were lost or accidently destroyed. Only nine researchers (24.32 percent) sent copies of their data.

Sharing ideas can be expensive. The time lag between submission of a paper to a journal and its actual publication can sometimes be one to two years. Therefore, conferences, congresses, and informal meetings have become major channels for exchanging new ideas (especially the so-called "invisible college" [Crane 1972[11]], which constitutes a network for the informal exchange of information[12]). To participate in this international network typically means using the Internet, but it also means expensive trips abroad. In most cases, young researchers cannot afford these trips. Their ability to be part of meaningful scientific networks may thus be severely limited, and sharing theories, data, and findings informally outside of journals may become an activity of very specific groups of researchers. The Internet makes such exchanges much easier, but still this fascinating means of communication has not replaced conferences.

The third Mertonian norm is organized skepticism, meaning institutionalized encouragement of researchers to criticize each other. Skepticism, however, is not a simple matter. An older, well-established, and secure researcher would feel much more comfortable criticizing a young, junior faculty member than the other way around. A junior faculty member in a department of statistics told one of the authors of this book[13] that he attended several lectures given by a specific senior faculty member. That person made some obvious mistakes, but neither our informant nor any of the other junior faculty members present commented on the mistakes, since none of them wanted to antagonize the lecturer, fearing that when time for tenure came around the senior professor would vote against them. In 1980 a new book on witchcraft was sent to Ben-Yehuda for review by a sociological journal. Ben-Yehuda did not like the book, and the review was, in general, negative. After the review was published, a senior sociologist in Ben-Yehuda's department commented to him that he should not have done that. He said that since Ben-Yehuda did not yet have tenure, it could be dangerous to antagonize other scholars. His advice was, of course, well-intended and based on real concern, but it did reflect that skepticism is a complex problem. Questions of who has enough freedom and security to criticize, and how and when, are crucial.

In another area, though, the norm of organized skepticism is perhaps the only norm that is adhered to in extremes. While this norm, in theory, is an invitation to make constructive criticisms without paying heed to personalities, in reality it frequently inspires and covers up for blatant aggression, condescension, and humiliation. Researchers who send a paper to a journal usually get it back with the anonymous referee's comments (who, in many cases, received the paper without the name of the author as well). This in itself could be taken to indicate the concern researchers may have in confronting each other openly, and it legitimizes hostility and aggression. It provides orthodox researchers with an excellent opportunity to kill new ideas (or vice versa; young researchers can use anonymity to delegitimize older ideas). In fact, because social networks exist among researchers, reviewers may have a pretty good idea whose paper or book they are reviewing. Moreover, since editors give much weight to referees' opinions, the researcher who submits a paper is virtually at the referee's mercy, as in many cases it is useless to respond to, or argue with, the referee's comments. Furthermore, referees can be chosen so that the editor can increase or decrease the probability of a favorable or unfavorable review. A similar logic, by the way, is valid for the review of research proposals, except that anonymity is not often used. Most researchers who have had to cope with referees' comments know how agonizing and demanding this process can be. In some cases the comments can be interesting, helpful, and constructive, while in others the comments are pointless, opinionated, aggressive, and useless. Such comments as "this is not my area, but . . ." or "this is unpersuasive" or the more common practice of giving many compliments and then, after a "but," following with a scathing and delegitimizing criticism are common occurrences. In fact, author Hank Campbell warned in 2014 that scientific credibility is harmed by what he views as the "corruption of Peer Review."[14]

Examples abound. A social scientist told one of the authors[15] about his research analyzing power interactions between therapists and patients and about a paper he sent on that topic to a specific journal. The journal did not accept the paper because one of the referees wrote, among other things, that psychoanalysts were not involved in power interactions and that "we psychoanalysts invite our patients to cooperate with us, but they need not believe any word we say." Another sociologist sent a paper to one of sociology's most prestigious journals. The paper was rejected because it did not use what the referees called "advanced statistical methods," as if no argument could be made without resorting to such methods. In this case, the use of such methods was not called for, as the sociologist had constructed her argument without the need for techniques such as regression analysis or correlation matrices.[16] Later the very same paper was sent to an equally prestigious journal of the American Psychological Association, received favorable reviews, and was published. When in the late 1930s, Grote Reber submitted his paper on radio astronomy to the *Astrophysical Journal* (the first time a paper reporting on radio emissions from the sky was submitted to an astronomical journal), the referees flatly rejected it because the reported findings were considered to be unbelievable.

This state of affairs, especially in the social sciences, could encourage contributors to overstate their cases. Researchers could underplay or manipulate data in a way that would be hard, if not impossible, to detect so that referees would allow them to pass. While such practice as overstating (or exaggeration) may not be seen by some as fraud, neither do they represent a scientific truth. The anonymous editor-mediated and determined interaction between author and referee is certainly not reminiscent of a free and open exchange of views. In such an interaction, power flows in one direction only. Under such conditions one can expect a certain degree of distortion and bias. Moreover, competition for publication is so fierce that it is possible that even one critical and negative review may condemn a paper (or a book, or a research proposal) to be rejected.

The fourth Mertonian norm is disinterestedness, meaning that a scientist's research should not be guided by expectations for personal profits and rewards. In reality, this norm cannot be maintained. Researchers derive two types of rewards from their work: personal-psychological and personal-economical. In some cases the two are interconnected. The personal-psychological rewards are mostly in the form of prestige, recognition, international reputation, respect, and the bestowal of various honors. Researchers' self-esteem can be high, and many bask in glory and respect. These benefits are certainly a far cry from disinterestedness. Especially since the latter half of the twentieth century, scientists are often consulted by high government officials and participate in top-level policy discussions on national, international, social, medical, and political issues. Many researchers appear on radio and television and enjoy popular admiration. There can be little doubt that Carl Sagan, through his 1981 television series *Cosmos*, earned himself a worldwide reputation (not to mention income). The scientist is perceived as the wise person who reveals the ultimate truth and unravels the secrets of our physical and social universe. The highest award any scientist can attain, the Nobel Prize, is very personalized. It is awarded personally to a specific scientist and not, for example, to an idea or invention.

The other element of reward is personal-economic. From the very nature of science's structure, researchers have a profound interest in the success of their ideas. Many researchers are motivated by wanting to solve a problem, push forward in the quest for knowledge, or for the cure for debilitating and lethal diseases. Nevertheless, they wouldn't be human if they didn't bask in the respect and glory that breakthrough discoveries yield. In turn, their achievements lead to better jobs, more grants, and more power and influence. Recognized, well-established researchers usually enjoy economic security and a moderate to luxurious lifestyle. Many travel throughout the world comfortably. Researchers have access to research grants that might further bolster their already comfortable salaries. Thus we find that researchers are attracted to fields where research is heavily funded, such as cancer, energy, law enforcement, gerontology, drugs and alcohol, and mental health. The list of fields changes with time and shifting priorities. Receiving money from research grants also means building up small "empires" of equipment, institutes, laboratories, manpower, and jobs. In other words, specific scientists can become powerful by virtue of their ability to provide jobs, salaries, and career opportunities. Aside from wishing to create new knowledge, interests may be propelled by personal motivation for recognition, fame, and self-esteem, or pure economic motivation, or all of the above. Quite a few of the rewards (e.g., the Nobel Prize) carry a hefty sum of money. In short, disinterestedness does not appear to be a realistic norm.

The Mertonian norms thus provide a weak defense against fraud because it is virtually impossible to fully conform to them. In fact, philosopher Warren Schmaus (1983) argues that fraudulent researchers violate common moral codes and not only codes specific to science. He thus places fraudulent researchers on the same pedestal as common fraudsters and con artists. Philosopher C. J. List (1985), on the other hand, suggests that we need to view fraud in research as a practice aimed at increasing the probability of success in acquiring rewards.

Replication and Reviewing

Two important control mechanisms against fraud and deviance in science are replication and refereeing. Of these two, replication is probably the single most important safeguard against deviance in science. "Reproducibility or replicability of experiments is the core tenet of the scientific method" (Katavić 2014). Replication is the essence of the scientific spirit. Irreproducible research[17] is problematic and in some disciplines is simply considered nonscientific. Unfortunately, replication itself is not without problems.

First, replication has not been considered prestigious or interesting work for researchers who have been trained to innovate. Few researchers would like to be permanently identified as "replicators" (Weinstein 1979). Furthermore, full replication is not always possible because in many cases not enough information is available about the original work. When sharing information becomes problematic, so does replication. Indeed, a number of researchers[18] have pointed out that a significant portion of published papers in psychology could not be replicated or verified because the original data were conveniently "lost" or "unavailable." In fact, a 2015 study reported in *Science* stated, "We conducted replications of 100 experimental and correlational studies published in three psychology journals using high-powered designs and original materials when available." The study[19] concluded that "[a] large portion of replications produced weaker evidence for the original findings despite using materials provided by the original authors."[20] This situation is aggravated by the fact that in many disciplines no institutionalized mechanism exists that keeps files or catalogs original data. In many cases, it is the researcher who is vested with the responsibility of keeping the data. In addition, granting institutes (as well as researchers) are reluctant to provide funds (or time and effort) for replication alone, unless the replication itself is significant in some important way. Even when one suspects fraud, if the replication (or checking the original data) would take a long time, it is quite possible that no one would be willing to spend the time to check it. If the suspected fraud was found not to have occurred, then the examiner-replicator would have spent much time on retests that would not even be publishable (not to mention the embarrassment to whoever raised the suspicion). In one case of suspected fraud (Colin 1984), a commission that worked hard found "no compelling evidence" substantiating the suspicion. Lack of full replication was one of the major reasons for this conclusion and the commission "urged" the experimenters to repeat their experiments.

In some disciplines replications are not only considered of secondary importance, but they are also problematic. Replication is most applicable to experimental disciplines, which lend themselves well to empirical experiments. It is not possible to replicate ideas and theories. Additionally, the use of methodologies such as surveys, participant observations, and ethnographies virtually guarantees that the results will not be replicable. Such sciences as sociology, history, and anthropology focus on interpretations. Observations, which are made at a specific time and place, are unique and not necessarily replicable, such as a study by an anthropologist who studies a remote tribe. In many such cases, even if two observations are made that have different results, the researchers can have legitimate explanations for the differences, since the nature of the observed phenomenon itself can (and does) change. For those who cynically want to utilize this fact to commit fraud, this could be a gold mine.

Replication, therefore, can only be considered a good safeguard against fraud in a few specific disciplines and fields. We learned from a famous survey researcher that if someone is sophisticated and well-acquainted with survey analysis techniques, she or he can manipulate questionnaires (by "playing" with how questions are phrased and distributed) in a way that could make the detection of fraud a daunting task.

Survey techniques, in this case, include all types of questionnaires. In 1976, when one of the authors of this book worked as a therapist in a methadone-maintenance program in Chicago, a few psychology students from Northwestern University asked to test some patients on the MMPI,[21] which is a long, complicated personality test. Some of the patients agreed to fill in the full version of the test. Weeks later it was reported that most of the patients who were tested showed schizophrenic traits. This finding had all the therapists in the program puzzled. On the one hand, psychiatrist Ronald D. Laing and others have maintained that drug addiction usually camouflages hidden schizophrenia, and the test results strongly supported this theory (with very specific clinical implications). On the other hand, most of the therapists who knew the tested patients and had previous clinical experience with schizophrenia could not concur with the findings. Checking this issue with the patients revealed that many of them had tired of the long test and took psychoactive substances, filling in parts of the test when they were under the influence of drugs. The testers should have detected this but had failed to do so. This illustrates that when data are manipulated at the basic level of actual collection, it can be extremely difficult to detect the bias.

Anthropologists who spend time among so-called "primitive tribes" sometimes witness what to them looks like bizarre events and patterns of behavior. If there were any suspicions about their trustworthiness and credibility, it would be exceedingly difficult to accept many of their observations. The case of the Don Juan materials of Carlos Castaneda,[22] an anthropologist who graduated from UCLA (Hooper 1978; de Mille 1980) demonstrates this point. There are those who argue that Castaneda reported nonexistent observations and merely wrote good-quality fantasy. Author Richard De Mille was apparently convinced that Castaneda perpetrated a literary fraud and told Hooper (1978, 29) that he "respects Castaneda as being both talented and shrewd, both in hoaxing the UCLA Anthropology Department and the reading public, as well as knowing the signs of the literary times . . . [and as] a 'trendy' writer."[23] In a similar fashion Arens (1979) states that there is no good evidence that cannibalism has been practiced regularly by any tribe and that the custom has probably never existed. This could imply that generations of anthropologists reported observations they never made and that were based on uncorroborated secondary sources.[24] In interviews conducted by one of the authors of this book with sociologists and anthropologists in the early 1980s, some interviewees admitted that as a "career line" they chose areas of specialization that no one could possibly replicate or examine, and so, under the convenient guise of "originality," they could discard "inconvenient" data, bias and twist reports, and get a great deal published.

Like replication, the peer review system employed by most journals is intended to ensure that the arguments and claims of researchers are based on solid research practices. In theory, reviewers may—at least sometimes—be able to detect suspicious work. In reality, however, few fraudulent procedures have been detected by reviewers. Our database indicates that only nine cases of fraud were detected by reviewers (about 1.2 percent!). Of these, two cases involved a reviewer who suspected plagiarism (in 2000) and a reviewer who suspected fabrication (in 1994).[25] This is a very small number of detections by reviewers.

One dramatic case of detection by a reviewer involved Stephen Pennycook and his team at a prestigious U.S. Department of Energy laboratory. Pennycook had submitted a paper to *Nature Physics* on the use of electron microscopes to analyze materials at the atomic scale. A reviewer for the journal, however, suspected that the researcher had manipulated the data. Confronted with these allegations, researchers at the laboratory admitted in 2006 to misrepresenting key data.[26] Another case took place in 1996 when a reviewer (for *Oncogene*) "spotted something odd in a paper about aberrant proteins created by rearranged genes" on a chromosome. This suspicion spurred an investigation, the result of which was that "Francis Collins, head of NIH's human genome project,[27] has informed colleagues that a junior researcher in his lab faked data in five papers Collins coauthored" (Marshall 1996, 908).[28]

In most cases (especially recent ones, see chapter 2), fraud was detected not by reviewers but by close associates or collaborators of the deviant researcher. The reason that reviewers find it difficult to detect fraud is simply because screening/reading a paper is neither a real full-scale replication nor a lie-detecting device. If what is written makes sense and does not clash severely with existing knowledge, then the probability that it will arouse the reader's suspicion is low. Wharton's marketing specialist Scott J. Armstrong (1982) analyzed what is published in scholarly journals and arrived at the conclusion that the chances of a paper being accepted for publication are tremendously increased if the paper is about an unimportant topic, is in agreement with existing beliefs, uses convoluted methods, and is written in an unclear way. Armstrong states that the probability of papers with "surprising" results being published is very low. Familiarity with these facts can be used by deviant researchers who choose to walk the path of fraud.

The following case[29] from the 1980s exemplifies the point. A team of three physicians in a well-known medical school received grant money to do a survey on the epidemiology of a lethal, although relatively rare, form of cancer. The team started the work, but after collecting data in about 50 percent of the cases two of the team members left for new jobs, leaving only one member to carry on. The work stopped for a few months, but then the granting institute started demanding reports. The physician still at the medical school contacted the other two members of the team, and together they forged the results. They doubled the results they had so they appeared to have data on 100 percent of the cases, but they found they got some "funny" results. So they used a table of random numbers to introduce some "random errors" into the original data. Then they checked the literature and found four old papers on the topic, all suggesting similar epidemiological dispersions and courses of the disease. The results the team had did not fit the literature, so they changed the data (by changing the actual numbers entered into the computer tapes) until they got results that were consistent with the literature. This fraudulent paper was presented at a scientific meeting and published in a well-known, prestigious, and refereed medical journal. Obviously, the researchers here took all measures necessary to cover their tracks, and it was virtually impossible to detect the fraud. The whole act of falsifying and fabricating data took them about ten days. Furthermore, since their results were consistent with the literature it was unlikely that anyone would want to replicate the work. Clearly, in this case no reviewer detected the fraud, and, indeed, how could it be detected by any reviewer? In the March 2007 case of Rebecca Uzelmeier[30] it was revealed that a similar technique of doubling data was used. If indeed careful doubling of cases is a mode of operation practiced commonly by fraudulent researchers, then the obvious difficulties of detecting it may suggest that a more prevalent problem exists.

Discussing fraud in cancer research brings up two famous cases of falsifying cancer patients' data. In 1993, Roger Poisson was found to have falsified data about breast cancer. He "was stripped of his post as head of cancer research at the Saint-Luc Hospital"[31] fined, reprimanded, and barred from various activities including being a principal researcher in medical research.[32] The other case involved Norwegian medical researcher Jon Sudbo, an oncology researcher who falsified data on about nine hundred patients from at least 1993 until 2006. Following a committee investigation, his authorizations as a physician and a dentist were suspended.[33] The editor of the prestigious *Lancet* is quoted as describing this as "the biggest scientific fraud conducted by a single researcher ever."[34]

We concur with Broad's conclusion in his 1981 survey of deviant researchers that replication and duplication of results is not a major factor in discovering fraud. Peer review is also not a strong safeguard against fraud. In fact, and based on two cases of recent suspected fraud, the June 14, 2014, issue of the *Economist* suggested that an open review on the Internet may be an improvement over a confidential peer review.[35]

Another problem with replication, or close examination of original data, is that sometimes existing knowledge or technology cannot detect fraud.

One most famous case is probably that of the Piltdown Man forgery by Charles Dawson.[36] In 1912, Charles Dawson, an amateur archeologist, came to Smith Woodward, keeper of geology of the Natural History Museum in London, with some human bones and a skull. Dawson claimed the bones came from Piltdown in Sussex and were found while excavating a pathway. These skeletal remains suggested the existence of a previously unknown human species.[37] While skepticism was raised, debates about this find raged. As time passed, the finding, named "Piltdown Man," became more and more anomalous and suspicious. More accumulated knowledge as well as more effective examinations made it possible eventually to declare the Piltdown Man a forgery (Weiner 1980). A similar but more modern case occurred in 2005 when it was publicized that an investigative panel at the University of Frankfurt had concluded that the director of its Institute of Anthropology, Reiner Protsch von Zieten, falsified data, plagiarized, and attempted to sell a collection of ape skulls owned by the university.[38]

Even modern technology, however, cannot always prove fraud. Sociologist H. Zuckerman (1977, 105–6) describes five examples of suspected fraud that are hard to establish as such, even with modern knowledge and technology. Of these five examples perhaps the most famous is the one mentioned earlier that involved the suspicions that Gregor Mendel may have falsified his statistics.

Apprehensions and Sanctions

While there are many motivations for committing fraud, the likelihood that a deviant act will take place depends, in principle, on two factors: the probability that the deviant will be caught and the severity of the sanction. Only where both of these factors are high can effective prevention be achieved.[39] The fact that most research is not replicated, and that replication may sometimes be problematic in itself, means that the probability that fraud will be detected is not high. Colleagues who are at the front line of being able to suspect or even detect fraud do not always feel comfortable to come forward with such unpleasant "news." The complicated and elaborate division of labor in science, which leads to a high level of specialization, may make the detection of fraud even more remote. Chances for replication are higher in meaningful research efforts or in so-called "breakthrough" research, and then they are done intensively. Zuckerman (1977, 95) pointed out that "the more consequential the scientific result, the more immediately efforts are mounted to reproduce it and thus, the greater the probability that error or deception can be detected." A good example involves the case of Marc Spector, who committed fraud on a breakthrough discovery that was certain to be carefully scrutinized and replicated as soon as possible.[40] Alas, most research is not of this nature. One interesting exception may be cases of fraud in specific disciplines in the public arena, in both digital and paper media, and, consequently, increased pressures on and efforts by members of such disciplines to make more efforts to replicate. In recent years, psychology may provide such an example.[41]

Furthermore, while the norms of organized skepticism and communality imply that each researcher is a potential police officer evaluating his or her colleagues' work, in reality this is not so. When police officers or detectives suspect that a crime has been committed, they are obligated to report or act and they have clear procedures and protocols for the detectives to follow, using both investigative technologies (legal, forensic, criminal) and institutions (police units, legal departments, international police support). If a researcher suspects that a colleague's work is fraudulent, how is that researcher supposed to pursue the suspicion? Who to address? What investigative tactics and technologies to use? When one is acquainted with the different known cases of deceit in research, it becomes apparent that the problem of detecting, reporting, and handling the suspicion is too often an ambiguous, complicated, and painful process. There is no international binding protocol of how to proceed when suspicions of misconduct are raised. To this one has to add that the question of who (or what) exactly is victimized by fraud in research (and therefore who should complain and when) has not always been satisfactorily resolved. A few illustrations should suffice. In two separate cases, we learned that colleagues in one discipline raised suspicions to both chairs of departments and dean, that they suspected colleagues in other departments to have been involved in fraudulent research. When asked whether they were willing to file complaints, the answer in both cases was negative. Whose role or duty was it then to file a complaint? To make a comparison, if chairs, deans, or other employees in an Israeli university hear about sexual harassment or abuse, it is their duty to report it. While it is reasonable to assume that honest researchers would complain about fraud they observed because it is their professional obligation and because it offends what their communal ethos is, it is not always the case. In another case, the person who filed a complaint demanded that anonymity be kept. Examining ORI's website reveals that the way cases of fraud were discovered, or, in other words, who blew the whistle, is seldom disclosed. We thus suspect that, overall, the probability of being caught is not very high. Obviously, the likelihood of being caught if the fraud is committed in what is breakthrough research is much higher.

The other factor to examine is the punishment for deviant researchers. For a long time, the typical process used when fraud was suspected was for the institution involved to appoint an investigative committee. Typical sanctions when these committees found misconduct were mild and ranged from warnings and reprimands to suspensions of research-related activities and revoking of licenses to practice. Deviant researchers typically have not been tried in courts and have not been subject to prison sentences, probation, or fines. The accused in these cases did not have any rights (or obligations) nor were they given a strong, formal, public way to defend themselves (or get a professional defense). Since the investigations are usually carried out within the concerned institutions, they are difficult for outside scholars to document and thus to conduct full-scale studies of this aspect of fraud cases. We suspect that in many cases the deviant researcher's act has not become a matter of public knowledge. The superiors of deviant researchers are never overanxious or too eager to disclose the case, for it might damage their own, and their institution's, prestige and credibility. If a disclosure is unavoidable, however, then these very same managers will typically show an eagerness to cooperate because this will show that their institution has high moral standards and demands honesty from its scientists. Thus if the case did not involve blatant fraud and fabrication on a large scale, the most that usually happened was that the deviant researcher lost his or her job. While this is a serious sanction, it hardly compares with punishments to which other white-collar offenders of organizational misconduct or fraud who were brought to justice are subjected. Very few fraudulent researchers were brought to court and even fewer were sent to jail. We do not have accurate comparative data for how many white-collar offenders or fraudulent researchers got away after some negotiations with their employers without any procedures in court. The "dark figure"[42] in both cases is probably impressive, as in both cases employers probably tend to prefer a quiet settlement rather than expose the embarrassment to their reputation and the legal costs involved in a criminal justice procedure.

Both Weinstein (1979) and Zuckerman (1977) have pointed out that punishment—up to the time of their work—was usually not very harsh. The scientific institution in which the deviant act occurred was usually not quick to admit it (i.e., Broad 1982). Four cases we described earlier—Darsee, Strauss, Long, and Summerlin—provide good examples. In the Darsee[43] affair (Braunwald 1992; Broad 1982), for example, even though Prof. James Darsee of the Harvard Medical School admitted that he falsified data and was stripped of his Harvard position, he nevertheless continued to work at the laboratory, publish papers, and appear at conferences. Only in February 1983 did the National Institutes of Health (USA) announce that he would be barred from receiving federal funds and contracts for a period of ten years (Wallis 1983). This, one has to notice, was an unusually harsh punishment at that time. Furthermore, Harvard University was asked to return $122,371 to NIH. Darsee, however, was in 1983 in the first year of a two-year fellowship in critical care medicine at Ellis Hospital in Schenectady, New York.

The Darsee case is particularly significant because it created quite a few responses and raised issues concerning the role and responsibilities of journals as well as the effectiveness of peer reviews.[44] The discussions about these issues were not limited to the Darsee case, and they addressed the above issues in a few journals in more general terms.

In another previously described case,[45] Marc Strauss, a researcher from Boston University, was accused in 1978 of submitting falsified reports. He insisted that he was the victim of a conspiracy planned and executed by his staff, but nevertheless he quit. It did not take him long, however, to reestablish himself and to submit a research proposal to the National Cancer Institute (NCI), which won him a $1.32 million grant.

John Long of Harvard Medical School, who forged data on cancer cells between 1978 and 1980,[46] later found a comfortable position. Summerlin, who was caught putting ink on tissue grafts of mice to "prove" a theory, was offered a long period of medical leave on full salary ($40,000!) to enable him to rest and obtain professional care (Culliton 1974, 1155).

While some of these relatively lenient responses may continue to the present, recent years have witnessed a hardening of responses reflecting changes to this pattern. Some fraudulent researchers were given suspended prison sentences and even sent to jail (e.g., Tufts University specialist and leader in managing pain Scott S. Reuben, who falsified data, or UK scientist Steven Eaton, who cooked experiments). This change seems to have occurred on the grounds of a new and more severe moral or ethical considerations about fraud, because more stakeholders became involved (e.g., journalists, granting institutes), and because some of the cases were viewed by granting institutes through the perspective of money. Since fraudulent research was frequently conducted using research funds, granting institutes could sue both the fraudulent researcher and the institution he or she worked for, which they did. This practice reminds one, obviously, that in past famous criminal cases suspected criminals were not brought to trial on charges of, say, suspected murders but for more minor offenses, such as income tax evasion. Al Capone is one glaring example. Nevertheless, this prima facie technical issue did not occur in a vacuum. The practice of using financial considerations to nail down fraudulent researchers could have been used years ago, but was not. This change in priorities and form of enforcement undoubtedly reflects more serious organizational changes in the moral and ethical bases of how to conduct research. Moreover, as more media attention was given to cases of fraud in research, more pressure was applied to process such cases properly, that is like any other case of fraud, and it became more difficult not to disclose such cases to the public.

The combination of a low probability of detection of fraud in research and relatively light punishments provided, in theory, a fertile context for deviance. But the perspective began to change from viewing such cases not only as a personal problem but as an organizational issue.[47] As stated above, in cases of fraud, granting agencies may demand that the organization where the deviant researcher worked return the research money (e.g., the case involving Darsee). In the long run this practice could mean that cases of fraud in research would not be dealt with on the individual-deviant level only, but that relevant organizations would have to find better, more efficient ways to detect and deal with deviant researchers simply in order to cut down their loss of both research money and prestige. It also means that the track on which processing fraudulent researchers would roll would be the economic path. Moreover, if abusing funds in this way would be referred to as "theft," then criminal charges, convictions, and much harsher penalties could be expected.

That there are organizational elements and forces that are conducive to researchers committing deviant acts is not enough in itself to explain the deviant acts. The element that is missing is motivational: Why should researchers commit deviant acts at all?

Motivations Favoring Fraud in Research

To state that deviant researchers have "something wrong" in their personality, socialization, or any other facet of their psyche or professional upbringing (that is, some variance of the "bad apple" theory) cannot explain their deviance. Aside from cases where fraudulent procedures were used due to some belief, cases we have indicate that many deviants knew fairly well that they were not supposed to do what they did, and in most cases they did not seem to be psychotic or otherwise disturbed. We must, therefore, look somewhere else for the motivations to deviate.

After the expansion of Western economies after World War II, there was an expansion of scientific establishments in the West, such as universities and research institutes, that lasted until the early 1970s. Opportunities began to diminish during the second half of the 1970s, however, and this trend of ups and downs continued, intensified by the economic crisis that began in the first decade of the twenty-first century.[48] Through these decades the notion crystallized that mobility in most Western scientific establishments depended primarily on published papers, books, and research reports. Moreover, academic careers began to become dependent on such bibliometric measurements as impact factors, citations,[49] and H-Index.[50] This process of the measurement-based "industrialization" of academic life put additional pressures on those aspiring to an academic career, as it placed a heavy pressure on researchers to publish. Moreover, at first the slogan "publish or perish"[51] was a meaningful phrase, and quality of teaching became secondary. But later on, as the economic crunch on higher education increased and students became increasingly defined as "customers," the quality of teaching became an important issue, which exerted additional pressures. Nevertheless, those who could write fast and had good contacts with editors and publishers flourished, and others agonized. When the necessity of producing a lot of work quickly is being impressed upon a researcher who very badly wants an academic career, recognition, fame, and security (tenure), a motivation to deviate is potentially created. In 1982 Broad noted that there had apparently been an increase in cases of fraud in science and quoted (482) Professor Ebert, former dean of Harvard Medical School, that "the enormous importance that is attached to the number of publications by committees that consider people for promotion . . . the pressure to produce, the pressure for publication" are probably the reasons for deviance in science.

Professor Ebert and Broad (1982) refer to the previously mentioned Darsee affair. In this bizarre case Darsee, from the Harvard Medical School, had nearly one hundred (!) published papers and abstracts to his name at the age of thirty-three. Apparently, this rate of publication was not enough for Darsee since in May 1981 he was caught falsifying an experiment. Investigations found that the original data for many of Darsee's experiments were unaccountably missing. Darsee himself confessed to only one case of fraud and denied any recollection of similar acts (Culliton 1983).[52]

Another previously described famous case of a young researcher in desperate need of publications is that of Vijay Soman of Yale University Medical School (Broad 1980b, 1980c). Soman admitted to falsifying work in a paper, and after an investigation, twelve papers involving Soman as a researcher were retracted (Broad 1982, 479). The Soman case, perhaps more than any other, reflects how pressures to publish created plagiarism, fudging, fraud, forgery, widespread destruction of laboratory data, and all else that could take place because no social-control mechanism worked effectively. Weinstein (1978, 13–14) reported on a 1977 case of a postdoctoral researcher in an institute of biochemistry in West Germany. This researcher spent almost two years measuring under various conditions the level of a certain chemical in neuroblastoma cells. Being under high pressure to produce, he published several papers reporting his results. After he left the institute some of his colleagues tried to replicate his work, with virtually no success. Consequently, he was asked by the institute to replicate his studies under close supervision, but he was unable to obtain the previously published results. Only then did he admit to "having invented the results of all his experiments." Likewise, in December 1982 a committee of the Mount Sinai School of Medicine found Dr. Joseph H. Cort responsible for the fabrication of data on drugs he claimed to have synthesized. Cort explained his fabrication by stating that he was under "a lot of pressure" to produce (Wade 1983, 3).

The organizational structural elements conducive to deviance in conjunction with the pressures just described can turn researchers into cynics. The main reason for this is that the ambiance in which they work emphasizes amounts of publications, places of publications, getting prestigious grants, networking with the right people, and going to the right conferences. Extension of certified knowledge is measured by the above-mentioned quantitative bibliometrics and indices. To achieve all these, one needs to excel in, and conform to, the existing paradigmatic ambiance. Creativity is not always congruent with this ambiance. For example, one of the authors of this book heard a young untenured colleague boasting that she managed to publish two papers in leading journals. Not a word was said about what these papers were all about. Cynicism could be the first step on the road to disenchantment, disbelief, and eventually deviance. If one needs to publish, and quickly, what could be more tempting than to commit (sophisticated) fraud that would be difficult to detect? This feeling of cynicism is probably magnified by other factors such as those noted by Weinstein (1978, 7): "Since one must publish to get grants, and promotion in many institutions hinges on the size of grants, publications and grants rather than discovery become the goals in the laboratory. . . . Scientists are tending to select research where money is available, such as cancer or energy research." Weinstein's assumption must be that scientists should be allowed to pursue research questions that they feel are important. The counter view is that since scientists are receiving grants from public funds, their research should reflect taxpayer's priorities as a form of accountability. The assumption in this view must be that that "the taxpayer," typically represented by elected politicians who are assumed to be fair and honest, knows what is required. A parallel view may exist regarding funds from foundations that are set to support particular goals (e.g., specific diseases).

The pressure from the grant-bestowing research institutes can also be such that it can help create tendencies toward deviance. John Long, an associate professor of pathology from Massachusetts General Hospital, admitted to the charge that he had forged data and attributed it to "pressure for a grant application"[53] (Wade 1981, 1023).

A different case (but similar in kind) occurred in May 1981, when Dr. Arnold Rincover of the Department of Psychology at the University of North Carolina was found by a faculty committee to be guilty of plagiarizing a student's thesis, almost word for word, in order to obtain a $723,000 grant (Wade 1983, 4).

Granting institutes, especially those that demand relatively "quick results," can help create a motivation to deviate. Fabricating or distorting data in such a direction that would show "success" can gain a specific researcher, institute, or laboratory considerable economical gains. In the Darsee fraud case it was stated that "the tremendous pressures to 'publish or perish' may be a factor. . . . These pressures have been exacerbated by the intense competition for limited federal research funds. 'Science is more expensive these days' says Albert H. Hastorf, Stanford's provost. 'You need a big grant or you are out of business'" (Wallis 1983). Organizations that deal with pressurized research grants, or are under strong pressure to "produce," may be good suspects for deviance. Thus, one should really not be surprised to find deviance in biomedical, biochemical, or pharmaceutical research (Silberner 1982).[54] In fact, Pontell, Jesilow, and Geis (1982) pointed out that in a specific medical-health-delivery system, the way the program was organized created incentives to commit fraud.

From conversations with colleagues, we have also learned of researchers who actively go to scientific conferences, contact fund executives to learn what the most recent "hot" area for funded research is, and make a quick switch to it. In many cases this appears to be simple scientific-economic entrepreneurship geared toward preserving someone's big laboratory (or research team), far from Merton's value of disinterestedness.

The structure of social mobility in science can also constitute a contributing factor. It is characterized by many relatively young scholars in the lower ranks (junior faculty) and only a few at the top (senior faculty). As our data analysis[55] indicates, the combination of age and peaks in cases of fraud points to hot spots where promotions typically take place. One peak corresponds to a time prior to tenure considerations and the second peak corresponds to a time before being promoted to full professorship. Young faculty members obviously desire upward mobility, but there are also intermediate positions of relatively young faculty members who are very close to becoming seniors but need a further push in publications to be granted that coveted status. One of the results of such situations, depending on the structure and pressures within a specific department, could be that this person might have his or her name printed first on research reports and papers she or he never wrote or that she or he had very little to do with. Although Zuckerman (1968) studied patterns of name ordering among authors of scientific papers, researching the elusive problem of "who did what study and under whose name it would appear?" is exceedingly difficult. Whatever the circumstances may be, if one's name appears as the first, last, or only author on a scientific work one did not conduct, one is getting credit for something one did not do. This is a fraudulent, deviant act that very much contradicts the ethos of science.

Trying to research this problem or assess its true magnitude is difficult, for who is going to openly and publicly admit such an act? Let us look at some cases recounted by our colleagues.[56]

Dr. K. H., a young pediatrician,[57] worked for a long time during 1980 and 1981 on a paper analyzing a specific disease. The work was done in a well-known, prestigious university medical school. When the paper was completed, she was instructed by the head of the department to put someone else's name as the first author. That person was due to be promoted and needed the publication. He had had absolutely nothing to do with the study. It is ironic that the results and analysis of the work were reported at a professional convention by that person, who had to spend a great deal of time with K. H. in order to acquaint himself with the study.

The second case is similar. Dr. T. I.,[58] the original researcher, was forced to appear second on the list of contributors for the same reason as K. H. This case took place in a clinical setting at a different medical school in the late 1970s.

A third case involves Dr. K. B.[59] who reported that in a certain university's department of political science, a paper appeared in a prestigious journal bearing the name of someone who had had nothing to do with the study the paper reported. That person was also due for promotion. In this case the original researcher's name appeared only in the acknowledgments.

The fourth case involves E. S.,[60] a now senior faculty member, in a psychology department of a large university. Prof. E. S. showed one of the authors of this book a book that was written in the late 1960s and bore the name of a certain author. The first name to be acknowledged is that of E. S.

According to E. S., he was the one who actually wrote the book, but the other person's name appeared because that person was due to be promoted and E. S. was promised that in return his career would be taken care of. Prof. E. S. specifically told one of the authors of this book that he would never forget the case and regrets very much having participated.

When these cases were discussed with two former chairmen and one current chairwoman in a medical school, their responses were almost identical:[61] "Why, this is a very common practice to help faculty" was the typical account. For a community that places such a strong emphasis on competition, integrity, and giving credit where it is due such a "common" practice does indeed seem bizarre. This practice may also work in different ways. For example, when a prominent scientist is able to get a grant based on his or her prestige, that scientist's name may appear as first author of a resulting paper even if that scientist had little to do with the research work involved.

Switches of name ordering on publications are also done by people who are close to promotion and want to "improve" their vitae. One of the authors of this book has heard of two cases where a person close to promotion, who collaborated with a more junior researcher, shifted the order of names and placed his name first although his contribution to the research was only secondary. Junior researchers seldom complain about having their names lower on the list, even when they feel strong personal discomfort, because they wish to avoid conflict with a senior collaborator and perhaps because they expect to be helped by such a procedure in the future, if needed.

Another relevant category includes a few examples involving the Nobel Prize. This prize, obviously, is a focal point for many researchers. The prestige and economic size of the prize are such that the norm of disinterestedness might become problematic.

The 1923 Nobel Prize in physiology was given to Macleod and Banting for discovering how to isolate insulin. Historian Michael Bliss (1982) has showed how Macleod (and others) twisted the truth about who should have gotten the prize. Certainly Macleod was not the ideal of a disinterested scientist. The thorny issue of who should have gotten more credit than who in this case is long and complex. While detailing this issue here is probably too complicated to be repeated, Bliss's book provides a richly fascinating account of this affair.

The more publicized example is reported by Zuckerman (1977, 170), Metz (1977) and others, and involves the discovery of a peculiar type of star, a pulsar. In September 1967, Jocelyn Bell-Burnell,[62] at the time a graduate student at Cambridge University, discovered a strange celestial radio source that pulsated regularly. This type of finding was considered impossible at the time. Bell-Burnell, however, pursued the matter and found three other such objects in different parts of the sky. The 1974 Nobel Prize in physics was awarded for this extraordinary discovery. The prestigious award, however, was not given to Bell-Burnell, but to Anthony Hewish, himself a radio astronomer and Bell-Burnell's graduate advisor. Although Bell-Burnell maintained that she got enough recognition (her name, for example, appeared on the paper announcing the discovery), others did not agree. Fred Hoyle, a famous and respected astronomer, publicly accused Hewish of deliberately understating Bell-Burnell's contribution (Zuckerman 1977). Motivated by the same feeling, the organizers of the Texas Symposium on Relativistic Astrophysics in 1976 invited Bell-Burnell to be the featured after-dinner speaker at their banquet. On that occasion, Tom Gould of Cornell University credited her with "perhaps the greatest single discovery in astronomy in this century."[63]

A similar case was reported by Harvey Fletcher (1982). Fletcher complained that when he was a graduate student, he suggested to Robert A. Millikan the idea of using oil drops instead of water drops in Millikan's famous experiment on the charge of the electron.[64] Furthermore, Fletcher claims that he wrote most of the 1910 famous paper that helped Millikan win a Nobel Prize. Millikan, however, took all the credit for himself.[65]

Finally, in 1975 Anne Sayre claimed that Rosalind Franklin was dishonestly, and systematically, deprived of her proper share of recognition for the discovery of the structure of DNA, for which James Watson, Francis Crick, and Maurice Wilkins were awarded the Nobel Prize in Physiology or Medicine in 1962. Later critical notes suggest that Sayre may have overstated her case.[66]

Getting credit for someone else's work reminds one of some old "traditions," probably still practiced in some places, where a senior scholar has one or many graduate students working for him or her for a long time. These students hardly—if ever—get proper credit and recognition for their work. The published work always carries the senior faculty members' name on it. Apparently, there are a few sociologists who have gained a reputation for this practice. In some instances, published papers based on graduate students' MA and/or PhD work carry as the first author the name of the faculty member who supervised that work. This can be done even though in most cases MA and PhD work is done as an independent work by the student. One of the authors of this book remembers vividly that both at Hebrew University and the University of Chicago, where that author was a student, students always warned each other not get into x's or y's project because it could be an "intellectually castrating experience." One was advised to get involved in these projects only if one really had no choice.

Taking formal credit for someone else's work brings us to the problem of plagiarism. While Zuckerman (1977, 117–19) pointed out how plagiarism violates scientific norms, Ben-David (1977, 250) indicated why plagiarism is easy to accomplish: "Scientific results, because they are specific and are independent of writing style, are easy to steal. Moreover, because of the frequency of genuinely independent simultaneous discoveries, it is difficult to detect plagiarism or, in the case of genuine multiple discovery, to assign property rights." That hundreds of journals exist in different languages and are not always accessible to all researchers makes the theft of ideas, and even whole papers, both possible and probable. The following case exemplifies this.

In 1977, a Jordanian student named Elias A. I. Alsabti came to the United States on a Jordanian fellowship, apparently to study medicine. Alsabti gave the impression that he was working hard on his PhD. In 1979 alone, he published more than sixty papers. While some of the papers were published in "junk" journals, others appeared in respectable publications. As it turned out, some of Alsabti's papers were proven to be pirated, word for word, from papers by others that had been published elsewhere. There is reason to believe that all of Alsabti's papers were pirated from other scholars. Broad (1980a), who vividly described this case, pointed out that Alsabti, apparently a wealthy and intelligent person, was motivated by a desire to become famous quickly. Alsabti's quite thorough knowledge of the academic system combined with his fraudulent methods of gaining entry into various U.S. schools and laboratories, enabled him to pursue his deviant career. He was shrewd enough to build his papers in a conventional way (adding a few fictional authors) and to be careful with the titles of the papers so as to make the detection of plagiarism more difficult. Alsabti, however, was not careful enough. Other scholars, especially those who worked with him, discovered some of his tricks and he was exposed.

Pressure to publish is not the only factor contributing to deviance. Another equally important factor is that one of the hallmarks of science is competition. Competition between researchers exists over discoveries, positions, grant money, recognition, and influence. Competition can cause deviant behavior proper or some very strange deviance-like behavior. Plagiarism is an obvious case. "Stealing" ideas is another one. For example, sociologist D. A. told one of the authors of this book that her team was preparing a research proposal for a large grant. She had a conference with a famous sociologist, Professor X, and discussed the proposal with him. Later on, her team's proposal was submitted and rejected. She was amazed to find that Professor X also submitted a similar proposal, improving on many of D.A.'s team's suggestions and projecting a lower cost. This proposal was funded. Another case was reported to one of the authors of this book in the early 1980s. An Israeli biochemist PhD student who was at a scientific conference met a senior female scientist. He was impressed with her work and spent almost a year in her U.S. laboratory. After this time he took all his notes, the skills and methods he had learned, and some of the synthesized materials and returned to Israel. This was done without the laboratory director's permission or knowledge. The work he did was in a very promising area of cancer research, and he was accepted back to an Israeli university. Apparently, no one there knew these details of his U.S. adventure. Shortly after arrival, he used the materials he had taken without permission from the U.S. laboratory to publish a few papers. The papers had as a first author the name of the Israeli laboratory chief, with the name of the PhD candidate next to it. The laboratory chief claimed not to have known that the results were "stolen" from somewhere else. A year later a conference on the same topic took place in Israel, and the PhD student opened it talking about "his" work. Unfortunately for him, the director of the U.S. lab was present (she was actually supposed to be the second speaker) and was shocked to realize what had happened. After many discussions, however, the American lab director decided to keep the whole thing quiet. The only result was that until 1983[67] the former PhD student did not get tenure despite many publications.

Another case surfaced when Stanford University began to investigate suspicions that Prof. Kenneth Melmon, chairman of the Department of Medicine at Stanford, "incorporated large chunks of copyrighted material from a book he helped edit into a chapter he wrote for another textbook. The material was used without attribution and apparently without permission" (Colin 1984, 35).

This issue has an interesting and instructive twist. Older, experienced, and well-established senior researchers often have young researchers working under them. An effective way to help a young scholar is to let his or her name appear as the first author of a paper. Zuckerman (1977, 170) reported that "[it] has been found for samples of American scientists that thoroughly established scientists, and in particular Nobel Laureates, often exercise noblesse oblige by ceding prime authorship to juniors whose careers are less secure." While Zuckerman used the positive label of "noblesse oblige," she nevertheless stated that this is a deviant procedure.

Ironically, taking credit for someone else's work can also have a boomerang effect. We saw this effect in some of the cases described earlier, and the following case illustrates it. In 1975, Robert J. Gullis published a lengthy work in biochemistry that summarized his PhD work done in Birmingham University, England. The name of Gullis's supervisor, Charles E. Rowe, appeared on the paper. The paper was considered important at the time. Unfortunately, in 1977 it was discovered by scientists at the Max Planck Institute for Biochemistry in Germany that Gullis, who was a postdoctoral student there, had fabricated data. While at the institute, Gullis also produced a few papers with the names of some German colleagues on them. When other researchers were unable to replicate the important work, Gullis was forced to admit that he had fabricated data (Broad and Wade 1982, 151–52).

Pirated work, plagiarism, and taking unjustified credit are all also related to the problem of "hired hand research" (Roth 1966). Hired hand researchers are people who actually write scientific papers but whose names do not appear on the final, published version of the work. Roth (1966) suggests that when these people realize that they have been hired to do somebody else's work, that they will receive no credit for the work, and that their suggestions may be ignored, there is a fairly good chance they will become disillusioned and careless with their work. Thus accuracy and precision could become secondary in importance, corners would be cut, and some faking, fudging, and fabrication of data could also happen.[68]

Another source of strain is focused on academics who might doubt their continuing ability to do worthwhile research. It might be that they have become administrators and feel the need to show that they are still capable of producing worthwhile research, or it might be because as they get older they experience increased anxiety about their ability to innovate. In both cases, such pressures may lead some to fabricate, plagiarize, or falsify research in order to assure themselves and others that their academic abilities have not evaporated.

ACCOUNTING FOR FRAUD

Accounts used by suspected research scientists are obviously interesting. Admitting wrongdoing and apologizing are highly symbolic expressions of remorse.[69] As noted in chapter 2, such stands have moral and legal implications. Unfortunately, such expressions are rare. As our data show, in 179 cases (51.6 percent of the cases on which we have information on this topic and 23.9 percent of the entire data base) the offenders admitted being involved in committing fraud. In less than 40 cases—about 22 percent of those admitting to have practiced fraud—did the offender apologize for committing fraud. That only about 24 percent of the entire database we have admitted wrongdoing and that only 40 of those apologized indicates how difficult cases of fraud can be. Typical accounts and patterns of admitting wrongdoing or regret include admitting the facts, stating that it was all a mistake, and that intention to commit fraud was not involved. Sometimes a claim was made that hiring a lawyer to help argue innocence was much too expensive.[70]

South Korean Hwang Woo-Suk[71] held a press conference in Seoul on November 24, 2005,

> in which he declared his intention of resigning from most of his official posts. This was after a committee found that he had fabricated stem cell research and personally coerced his female researchers to perform egg extraction procedure for his studies.[72] He also apologized for his actions. In the interview he said, 'I was blinded by work and my drive for achievement.' He denied coercing his female researchers into donating eggs and claimed that he found out about the situation only after it had occurred. He added that he had lied about the source of the eggs donated to protect the privacy of his female researchers, and that he was not aware of the Declaration of Helsinki, which clearly enumerates his actions as a breach of ethical conduct.[73]

Given the moral nature and impact of an apology, it is perhaps not surprising that "[a]fter the press conference, which was aired on all major South Korean television networks, most of the nation's media outlets, government ministries, and the public gave support to Hwang. Sympathy for Hwang poured out, resulting in an increase in the number of women who wanted to donate their eggs for Hwang's research."[74]

Two illustrative cases are, perhaps, at the other moral pole. The first case involves German chemist Guido Zadel of the University of Bonn. "Two years after he allegedly was caught trying to manipulate the results of a laboratory experiment based on his thesis, the University of Bonn withdrew the doctorate it had awarded [him]."[75] German correspondent Schiermeier reported that Zadel, "a former chemist at the University of Bonn, who has been accused of committing scientific fraud four years ago, is suing the *Land* (state) of Nordrhein-Westfalen for DM437,000 (US$242,000) compensation.

Zadel's lawyer claims that, as a direct result of the accusations, he has been unable to find a suitable post, and that the sum represents the amount a chemist of his age and qualifications could expect to earn in industry over four years. Zadel is also seeking DM200,000 in damages from a former co-worker who has said that he saw Zadel manipulating experiments. Zadel is challenging a decision by the university to withdraw his doctorate."[76]

The second case took place in 1997 when the Swedish Karolinska Institute distanced "itself from the work of one of its former researchers, Ulf Lönn, six months after an independent panel of investigation concluded that he had 'most probably' committed scientific fraud." But "the institute stops short of accusing Lönn formally of misconduct" and "Lönn believes that he has been legally 'cleared.' His lawyer . . . says that the investigation did not give the scientist 'a fair chance' to defend himself, and he may bring a court case."[77]

Our database has a number of quotations from various lawyers who were involved in various cases of suspected fraud. While at first we considered using some of these legal and technical accounts, we decided not to pursue this line. Our interest was to examine how suspected researchers and organizations respond to allegations of fraud and not to show how sharp or not so sharp lawyers are at representing and constructing various legal claims.

FAMOUS SCIENTISTS AND RESEARCH FRAUD

The Mendel case is only one instance of famous men of science who may have committed various acts of misconduct. Kinsey is another obvious suspect. Two other famous cases involve documented cases of suspected fraud by such eminent researchers as Nobel laureate Robert Millikan (discussed elsewhere) and fudging by English physicist and astronomer Isaac Newton (Westfall 1973) as well as Newton's falsifications in his priority dispute with Leibniz (Manuel 1968).

Some works[78] strongly suggest that eminent past scientists committed acts of colossal fraud. Gerald Geison's (1995) biography of Louis Pasteur reveals that this eminent French scientist did not give credit to a competitor whose technique he used to create a vaccine against anthrax, that there were cases of misrepresentation in his work that were aimed at raising funds and boosting his image, and that he performed experiments on human subjects without their consent. Traill's (1995) study suggests that at least some of Heinrich Schliemann's published work on his famous excavations in Troy and Mycenae were fraudulent.[79]

French astronomer Pierre-Simon, Marquis de Laplace was suspected of stealing ideas "outrageously, right and left, whenever he could," and German mathematician Gottfried Wilhelm von Leibniz was also suspected of similar acts. Italian physicist and astronomer Galileo Galilei was said to "shamelessly" have stolen ideas from German astronomer Johannes Kepler and others. Likewise, Graeco-Egyptian mathematician and astronomer Claudius Ptolemy, as well as Dalton, Lavoisier, and Pascal, may have all been involved, and possibly guilty of, some form of deceit in their work.[80] Author Jeffrey Masson (1984) has accused Freud of deliberately and intentionally ignoring some of his findings in order to pacify public opinion and be more acceptable to the medical establishment. In a similar fashion, Vanderbilt University sociologist George Becker (1984) accused Merton of misinterpreting and ignoring sources (to the point of faulty use of evidence), thus calling into question Merton's conclusions regarding the relationship between German pietism and the rise of science over the past four centuries.[81]

FRAUD IN RESEARCH AND THEORIES OF DEVIANCE

Fraud in research occurs because of at least two clusters of reasons. First, it becomes possible because social-control mechanisms in science are relatively weak. There is a fairly good chance, though, that much tighter social control, even if possible, might conflict with the nature of the scientific work itself, making it more bureaucratized and less open to innovations. Second, science and the scientific community possess structural and motivational factors that are conducive to committing acts of fraud. The pressures to deviate may have less to do with the researcher's personality, but are related to the very structure and organizational context of science, its goals, the way it functions, and the specific positions research scientists occupy.

Finding a sociology of deviance theory that can explain why or how a researcher becomes deviant is not a simple challenge. The prevalent labeling theory[82] of deviance, for example, could apply, but only partially. Fraud in research is committed in secret because it is in the researcher's interest to keep it hidden, and the detection rate is probably low. Nevertheless, once discovered the shaming and disrepute can be quite devastating, more so if it becomes public knowledge. While labeling theory is right that making research rules means the creation of conduct norms whose infraction constitutes deviance, it cannot explain persuasively why fraud in research occurred in the first place. For example, imagine a research scientist in a university setting under pressure to be promoted and does not have enough publications in what are considered top-notch journals. He or she then might ponder about falsifying publications or experiments, and the best possible strategy would probably be to publish a lot in good journals but not to "make waves." That is, reduce the risks of drawing attention, of having someone try to replicate your work, of starting a controversy, in short, of being caught and labeled with a stigma of a fraudster. Publishing much in what are considered to be second- or sometimes even third-rate, but viewed as good or OK,[83] journals can minimize or even avoid the chance of replication and the exposure of the researchers' deviant acts. Unlike the drug user who may indulge his habit in company or the embezzler who wants to flaunt money, successful fraudulent research will never be detected and will enable the deviant to lead a comfortable, interesting, and prestigious life.

Nor can one apply Sutherland's differential association theory[84] to fraud in science. While this theory explains how one can become a deviant, it states explicitly that deviance is a behavior learned through association with other deviants who provide the novice with the ideology and the technology of why and how to become deviant (e.g., the ideology of thieves and the "know-how" of stealing).[85] Deviant researchers, however, would rarely—if ever—associate themselves with other deviant researchers. Another group of theories, traced to Robert Merton, suggests that structural elements such as blocked opportunities, deprivation, and frustration can push potential deviants into actualizing the potential. Other theories search for causes of deviance in such personal attributes as defective personality, immaturity, or a lack of morality.[86] Nevertheless, research scientists in many cases do not come from deprived homes or slums,[87] nor are they psychotics suffering from various genetic, brain, or hormonal problems that may cause deviance, nor do they simply suffer from serious personality disorders or are otherwise crazy.

The various theories of deviance developed originally to explain deviances such as theft, prostitution, addiction, murder, and juvenile delinquency do not apply to deviant researchers. The most promising way to understand deviant scientists and interpret their acts is probably through the perspective of social control. Of the few control theories available,[88] some may help us sort out an answer to the puzzling question of researchers who choose to commit fraud.

We may want to begin with one interesting possible interpretation that relies on sociologist Donald Black's works (1983, 1984a, 1984b), which argue that we can view deviance as an expression of countermorality and as a pursuit of justice by the deviant. In our case, a person who may feel "locked in" demands to be creative, to publish, to teach, to get research grants, may resort to acts of intentional deviance—fraud—as a way to relieve the pressure and reach a desired goal of prestige that comes with a successful innovation or publication, while at the same time signaling (and justifying the act) that requests are too high, too demanding, and too oppressive. Outsmarting the "system" by resorting to fraud may have been a thought that crossed some fraudsters' minds. To a limited extent, Barber et al.'s 1973 study on using human subjects in medical experiments lends some support to Black's work. They point out that researchers "who are relative failures in this competition or who feel they have been unfairly treated are more likely to express permissive standards and manifest permissive behavior in the use of human subjects" (189). The limited support is because these researchers do not really want to "get even" with the system, but rather they look for a deviant easy way to get into the system. Overall, Barber et al.'s 1973 work focused on experimentation with human subjects and placed a strong emphasis on macro-institutional control approaches. They thus examined some of the structural elements involved in conducting research (e.g., review processes). For example, they pointed out that "medical schools have not been ethical leaders in respect to efforts to safeguard the rights and welfare of the human subjects of biomedical research" (177).

Another development in control theory was made by sociologists Michael Gottfredson and Travis Hirschi (1990), whose main explanatory concept is that deviance occurs due to low personal self-control.[89] Their theory, as well as Black's, attempts to explain how a researcher may move from point A, where he or she is nondeviant, to point B, where he or she is a bona fide deviant. Throughout this book we have examined various structural, organizational, and ethical mechanisms of social control in research. As we saw, not all work perfectly. Here we are more interested in suggesting what may possibly take place in a potential deviant researcher's mind as he or she may trot along the way from being nondeviant to being deviant. Aiding in our understanding and detailing this process is sociologist David

Matza's control theory (1964, 1969), which is perhaps the most appropriate to our subject and is discussed below.

In his attempt to explain how people become deviants, Matza takes on one of the most difficult dilemmas in the sociology of deviance: Are humans free agents, able to choose a deviant career, or are they passive subjects whose conduct is determined by forces over which they may have little, if any, control? Matza emphasized the process of *becoming* deviant, mapping the factors that facilitate deviance and taking into consideration human choice as well.

The last few pages of *Delinquency and Drift* (1964) lay the groundwork for Matza's theory that charts the process of a nondeviant person becoming a deviant. First, there is the concept of "drift," that is, what occurs when an individual becomes detached from a specific societal group or organization or from society's more general moral codes. For a drift situation to come about, two conditions must be met. The first is neutralization. By this Matza is referring to the process through which it becomes possible to neutralize the moral grip of society on the individual (i.e., denial of responsibility, denial of injury, denial of victimization, and condemnation of the condemners [Sykes and Matza, 1957]).[90] Crucially important moral codes that a potentially deviant researcher needs to neutralize are the trust of his colleagues and the organization within which he or she works as well as professional obligations, honesty, pride, and reputation. The second condition is what Matza called "subterranean convergence," an analytical concept that refers to the process through which conventional morality and culture mixes with, and gives support to, deviant morality. Some illustrations of subterranean convergence are the cases we have described of researchers who deviated because of belief in a specific theory or because of cultural and political factors intervening in the scientific process. Lysenko is a good illustration.

According to Matza, when subterranean convergence takes place together with neutralization, the probability of entering a drift situation becomes very high. Being in a drift situation, however, does not guarantee that deviance will actually happen. For that to occur, three more conditions must be met. The most important of these conditions is the motivating element, that which actually propels a person to deviate. This is what Matza titles "will." Will, however, has to be created, and its creation depends on two other conditions: preparation and desperation. Both conditions indicate specific feelings that, as sociologist Charles Frazier (1976) indicated, activate the will. When subterranean convergence couples with neutralization to produce a drift-like situation, and when preparation and desperation join together to activate will, deviance will take place. In *Becoming Deviant* (1969), Matza added society's signification of the deviant as a deviant. When societal reaction (labeling) takes place, it can lock the previously nondeviant researcher into the social role of a deviant. What this element may also imply is that it is possible that once a person self-identifies as one who can and does commit an infraction, more infractions will probably follow.[91] In his two books Matza thus interpreted the process of becoming deviant as an interaction of external, uncontrollable factors and the individual's free will.[92] In the case of fraudulent researchers whose acts become public knowledge, the societal reaction will have a significantly different role. We did not find significant empirical support for the notion that public negative reactions to fraudulent researchers (e.g. shaming) locks them into this role and they become recidivists. The likely reason for that is probably because identifying a fraudulent researcher provides such a blow to that researcher's career that recidivism is not a realistic course of action. Moreover, chances are that if they continue to work as researchers their work may perhaps be monitored more closely.

The previous sections that described the structural elements conducive to deviance in research and the motivational aspects of this deviance in organizations fit very nicely into this scheme. Deviant researchers are not born to be that way; rather they become deviants. A lonely researcher, hungry for publications, recognition, and rewards, in a stiff competition and in a desperate "need" for tenure can—relatively easily—enter into a situation of "drift," and hide it quite effectively (in some disciplines better than in others). Neutralization of guilt can also be provided (for example, fabricating results to support previous research so that the possible damage is thought to be minimal, or feeling that the system treated him or her unfairly). Together with the low probability of being caught and the perception that the punishment for those who are caught is not too severe, deviance can become almost irresistible to some. Moreover, and as we saw before, subterranean convergence may happen if one believes that all means are legitimate to achieve a goal (remember Blondlot and N-rays? Or Lysenko?). Once in a drift process, one is not too far from having the will, charged by desperation and preparation, to become deviant. The ease and speed with which fraud in research can be committed only reinforces this.

Moreover, a 2016 review in the science section of the British *Guardian* pointed out that pressures to publish can result in "[s]cientists incentivised to publish surprising results frequently in major journals, despite risk that such findings are likely to be wrong." The review relied on cognitive scientist Paul Smaldino's estimate that "[s]ociology, economics, climate science and ecology are other areas likely to be vulnerable to the propagation of bad practice."[93]

PREVALENCE OF FRAUD: BAD APPLES OR AN ICEBERG?

Zuckerman (1977, 107–8) noted the strange fact that science, which is devoted to the careful formulation and preservation of detailed records, had not introduced systematic procedures to record and analyze deviant behavior of researches. She observed that in 1977 science lacked a formal code of professional ethics and a body of pertinent law and that there was no organ whose job it was to detect, prosecute, and punish. Therefore, the basis for a systematic record keeping of deviant acts was altogether lacking. There was no scientific or other police-like force that chased suspected deviant researchers and scrutinized their records. No universal protocols for investigating suspected fraud existed and there was no organizational structure to help develop specialists in investigating fraud or misconduct. In a number of cases, scandals were not reported via the scientific communications network but became public through such mainstream newspapers as the *New York Times*, the *Boston Globe*, and other newspapers (Woolf 1981). *Science* and *Discover* were probably the major journals that tried to report on deviant scientists. The few reports which did exist were just that, journalistic reports. There were virtually no systematic scientific studies of fraud in science. Nor was there a systematic study of deviant researchers as a category.[94]

A few prominent cases of research fraud in the early 1980s, though, triggered action by both U.S. legislative and administrative organizations. Rules were enacted and organizations were created. Eventually, in May 1992, the Office of Research Integrity (ORI) was established. ORI tried to set rules, keep records, and get involved in suspected cases. Likewise, ethical codes became popular and research organizations practiced more training and control aimed at reducing the dangers of fraudulent research.

Nevertheless, and despite significant changes, current available information is still partial and lacking in many categories. The problem of the "dark figure"—how much we know about various manifestations of deviance vs. cases that are hidden—plagues this field. Empirically, it thus remains an open question as to how pervasive fraud is in research and how much gets exposed and punished.

In science, as in similar areas of deviant behavior (i.e., police corruption, embezzlement in banks), there is a bitter argument between those adhering to the "iceberg theory" (we are only seeing the tip of the problem while the large part of the problem is hidden from the public) and those adhering to the "bad apples" theory (there are only a few cases, but they make the problem appear bigger than what it is) (Kilbourne and Kilbourne 1983).

The critical discussion of the Mertonian idealized norms, of replication and refereeing, and of the structural and motivational aspects conducive to deviance in research all indicate that the prevalence of deceit in scientific research is probably much greater than we thought. The number of known and publicized cases is, indeed, relatively small, but coverage of this problem is biased. Many of the reports about fraud are in those areas where "hot" issues or "breakthrough" research are debated. Replications are more frequent in these areas and work is supervised and monitored more closely. Thus the disclosure of fraud cases should not be taken as an indication of the low prevalence of this type of deviance but rather as an indication of under what circumstances fraud gets publicized, who gets caught, and why.

We would—very much—like to believe that most scientific research is far from the type of work in which acts of deviance have been discovered. In many areas we probably have researchers who are doing their jobs honestly with a genuine passion for extending certified knowledge. Nevertheless, there are many areas where non-breakthrough research is being done. If a carefully planned act of fraud or deviance is being committed in one of these areas, the probability of its ever being detected is very slim. Also, we may expect deviance to be more prevalent in commercial-scientific areas such as pharmaceutical research because there is more at stake there and thus more corner cutting may take place.

The structural elements conducive to deviance in scientific research and the lack of scrutiny that most research receives suggest that the "iceberg theory" is probably closer to the truth than the "bad apple" theory.[95] Support for this conclusion came from British author Daniele Fanelli, who reported in 2009 about a meta-analysis of a large number of surveys. The results showed "a pooled weighted average of 1.97% . . . of scientists admitted to have fabricated, falsified or modified data or results at least once—a serious form of misconduct by any standard—and up to 33.7% admitted other questionable research practices. In surveys asking about the behavior of colleagues, admission rates were 14.12% . . . for falsification, and up to 72% for other questionable research practices."[96] Further support came in a 2012 disclosure that about 75 percent of 2,074 retracted papers in biomedical research were retracted because of misconduct.[97] In fact, Gøtzsche's 2013 book calls for more attention to the scale of fraud in biomedical science by exposing practices of fraudulent behavior of pharmaceutical industries. It also seems reasonable to assume that while fraud is probably more prevalent in research than most naïve people think, it is also probably less prevalent than in other areas, such as commerce, dealerships, consumer fraud, or con games.[98] Finally, the British *Spectator* published in October 2016 a review article that argued that many scientific papers cannot be trusted.[99]

FRAUD IN RESEARCH AND SIMILAR FORMS OF DEVIANCE

In chapter 1 we pointed out that fraud in research is not dissimilar to fraud in other areas such as police or banking, where the term "misconduct" is also used to describe the phenomenon. In this section, it is our goal to dive deeper into this issue.

Deviance in scientific research resembles professional deviance, white-collar criminality, and government deviance. Honest actors in these areas enjoy a perception of being serious, decent, and respectable and of having (and using) extensive power, as well as being involved from time to time in morality statements and negotiations, thus helping to shape mainstream social realities and definitions. Deviants in these areas disrespect all these attributes and abuse the trust and loyalty of their victims. Likewise, research scientists are perceived to be serious, honest, responsible, and respectable, and quite a few occupy positions of power, educating the public and influencing social, national, and international policies. Many times, such people are presented as model citizens, successful by all economic and social standards, bearers of conventional morality, and recipients of the benefits of being loyal, conforming members of the social system (i.e., Douglas and Waksler 1982; Thio 1978). Yet unlike honest researchers, fraudulent researchers display cynical contempt for the truth, whatever it might be. They use their knowledge of science to abuse science and to ruthlessly advance their careers, betraying the trust and loyalty invested in them.

There are a few forms of deviance and crime the definitions of which could fit fraud in research. Sociologist Richard Quinney (1964) characterized professional (or occupational) deviation as those forms of behavior that deviate from the norms of a specific occupation or profession, but which are not illegal. This characterization may fit some cases of deviant researchers and deviance in science. White-collar crime, on the other hand, would be characterized as professional deviance, but would be illegal. For example, while fudging and suppressing data could fall into the first category, plagiarism and fabrication of data could fall into the second category. Indeed, sociologist Gilbert Geis (1977) has suggested that white-collar criminality should include deviance in the professions. In fact, Green's work on white-collar criminality includes an entire chapter on fraud (2006, 148–60). Fraud in research can thus be also thought of as part of professional deviance.

Cases of deviance within police departments are also similar to the case of deviant researchers (Douglas and Waksler 1982, 353–59). Like the structure and the creation of motivation conducive to deviance in science, sociologist McCaghy (1976, 191) points out that "tendencies toward corruption are built into the law enforcement system."

There are a number of similarities between deviance in research and professional, government, and white-collar deviance. First is the problem of damage. Professional deviance, such as cases of pollution, unsafe machines or drugs, dangerous food, fraudulent research in pharmaceutics or unnecessary surgery, could all result in numerous victims. In the social sciences, fabricated or biased results on IQ tests could damage the image of a whole category of people. White-collar deviance can result in numerous victims and sometimes cause heavy economic damage. But the main damage of all these forms of deviance, and of deviance in research, lies elsewhere. White-collar crime clearly damages important social values and norms like trust and morale. Government deviance such as political corruption or illegal police activities (Thio 1978; Douglas and Waksler 1982) can breed contempt for the law, cynicism, and disrespect for social order (Thio 1978). In the same way, fraud in research can lead to distrust of science and may shatter the legitimacy of science itself. Indeed, Kilbourne and Kilbourne (1983, 133–154) pointed out that deviance in science can lead to a depletion of a vital US national resource, that is, science itself.

Second, the problem of evaluating the prevalence of deviance is similar. In all the aforementioned forms of deviance, it is in the interest of the non-deviating relevant group to show that it can detect deviance and punish the deviant. Like the organizations in which scientific research is conducted, however, these groups are not interested in finding too much deviance because it could undermine their own legitimacy.

Third, detection and punishment can also be problematic in these areas of deviance. It is difficult to detect deviants in the professions, government, corporations, and the sciences, not to mention high-level white-collar deviants. Many criminologists and sociologists of deviance seem to feel that, as in research, punishment for deviants in these areas has been—relatively speaking—easy. In fact, Thio (1978, 743) suggested that the above-mentioned forms of deviance should be called "profitable deviance" because "the individuals involved can illegally acquire a lot of money, power or influence with a relatively low risk of harmful consequences to themselves." Wheeler, Weisburd, and Bode (1982), however, present a different view. Their work points out (a) the complexity of punishment, as well as its predictability, and (b) that there are white-collar criminals that are sent to prison.[100]

Fourth, it appears that the essence of professional, government, and white-collar deviance is similar to deviance in research. All are focused around violations of, or deviation from, socially defined professional, business, or political roles (Edelhertz 1970; Conklin 1977).

Fifth, in all the aforementioned areas of deviance, actors who are otherwise considered respectable, have no past criminal record, and appear to be models of conventional morality commit the deviant acts.

Nevertheless, there are also at least two marked differences between these forms of deviance and fraud in research. First, white-collar criminals are usually after money or other economic gain. Fraudulent researchers, perhaps more like deviant government officials, are more after recognition and job mobility. Second, while the results of the deviance in the cases we have discussed can be severe—in practical, nonsymbolic terms—one has to recognize that the amount of damage done by fraud in such areas as anthropology, sociology, political science, communications, international relations, history, or literary studies can be negligible. Fraudulent work in these areas can result in a wrong interpretation, the damage of which need not be very high.

ACADEMIC POLITICS AND FRAUD

Suspicions and accusations of fraud in research can create quite a turmoil within departments and universities. Two cases that took place within Hebrew University illustrate this. One case took place in the faculty of medicine, where a physician was suspected of research fraud. As suspicions were raised and an investigation took place, the case reached newspapers.[101] The public could watch with amazement as various accusations were made regarding the suspicion and the behavior of the physician, accompanied by claims of improper behavior in this case by some top members of the university. Various (and sometimes conflicting) stakeholders expressed support for the suspected physician; others questioned the conduct of some Hebrew University officials. Eventually, the case was brought before a university disciplinary court. The physician was found guilty of research fraud and sanctions were applied.[102]

Another illustrative case involved a faculty member who was suspected by a colleague of research fraud. As in the previous case, a committee was appointed to investigate the suspicions, and as in the previous case, details of this case reached the media, where different—sometimes conflicting—accounts were given. In this case, the affair ended when the suspected faculty member retired from the university and no disciplinary action took place. As this affair was unfolding, different stakeholders from within and outside the university began to intervene. For example, supporters of the faculty member questioned the suspicions, the procedure, and the qualifications of the committee's members.

As these cases demonstrate, investigating suspicions of fraud can be a nasty affair. Such suspicions and investigations are immersed in political, organizational, and cultural contexts and may involve various stakeholders with hidden or explicit agendas whose response to a case may be influenced by those agendas rather than by the case itself. Some senior faculty members may have networks both within and outside their institutions that may provide an envelope of protection that may make a fair and useful investigation of any impropriety difficult. One faculty member used an instructive imagery. She stated that inner conflicts, likes, and dislikes, as well as networks of support or hostility may make some departments appear like a gas station with fumes in the air. In such places, an accusation against a faculty member can have the effect of igniting a match in that gas station.

CONCLUDING REMARKS: REACTIONS TO FRAUD IN RESEARCH—A QUESTION AND RHETORIC OF BOUNDARIES

The way the scientific community regards acts of fraud by its members is instructive. Obviously, this community, like any other organization, has a vested interest in minimizing publicity about deviant cases. Too many public disclosures of hoaxes, fabrications, and fraud could severely damage the credibility and prestige of the scientific community. Admitting that public, political, or cultural factors could bias scientific research could have much the same result. It is also very important for the scientific community, however, to expose cases of deviance and thus show to the public, as well as to its own members, that it has the ability to detect such cases, to be open about them, and to punish the relevant scientists.

It is predictable then that the reaction of the scientific community in its attempt to try and defend itself will exhibit a complex and not necessarily consistent pattern. One obvious form of defense is not to keep records of cases of deviance that have been reported. The ORI's activities do not support this, but then not every case of fraud is reported to the ORI and the not always detailed nature of ORI's reports does add to this observation. An even better illustration is the British Website that is supposed to reflect coping with fraud in research: UKRIO, or the UK Research Integrity Office. Completely unlike the ORI, this anemic Website provides zero information about real and actual cases or how they were processed but instead provides advice about ethics and integrity in research.

Representatives of the scientific community may also deny that there are organizational factors that contribute to and facilitate fraud in research. This situation is reflected in a double-edged position echoed by different spokespersons. One edge, probably the more common form of defense, is a mild to fierce denial of the pervasiveness of fraud in research.

The denial of the possibility that fraud in research is much more prevalent than previously assumed is perhaps best exemplified by Zuckerman's (1977, 98) conviction that "[a]lmost all those who have bothered to look conclude that seriously deviant behavior in science is rare." and then the amazing validating phrase, "scientists themselves repeatedly note this one fact." Zuckerman does not stop there and brings in a few quotations from different scholars to reinforce this view. Noteworthy is Merton's opinion that cases of fraud are virtually nonexistent in the annals of science. Zuckerman, however, is careful to point out that this view is impressionistic and not grounded in facts. She points to the absence of statistics on deviant behavior in science and—to her credit—comments that this situation has inevitably aroused speculations.[103] Thus, while she notes that there were cases of assumed deviant scientific acts by such distinguished scientists as Ptolemy, Galileo, Newton, Mendel, and Freud, she also hastes to contradict them. Let us examine a few other relevant illustrations.

Kevin McKean, who so vividly documented the Spector case, also supported the idea that fraud in research seldom occurs (1981, 23). He noted triumphantly that "whatever new facts emerge from the Spector case, it remains true that fraud in science is rare" (33). Author Isaac Asimov, in his 1979 short and informative "Alas, All Human," also gave the strong impression that fraud in research is unrepresentative and rare. Interestingly, while sociologist Patricia Woolf (1981, 9) noted that "scientists as a group are generally reluctant to acknowledge falsification of data or a pervasive problem and seem unwilling to take formal notice of this serious deviation from prescribed scientific norms," her paper does leave the reader with a distinct feeling that fraudulent researchers are truly a rare breed. And she quotes (10) Philip Handler, president of the National Academy of Sciences, who said in testimony on March 31, 1981, that the problem of falsification of research results "has been grossly exaggerated. . . . It is rather a relatively small matter which is generated in and is normally effectively managed by that smaller segment of the larger society which is the scientific community."

Wrapping, cloaking, and reframing the negative nature of fraud in scientific research can employ some inventive and more morally neutral terms such as "misconduct" or "irresponsible behavior" (Steneck 2006) and can even resort to inventive terms such as "intentional bias" (St. James-Roberts 1976). These terms tend to weaken and minimize perceptions of the nature and prevalence of fraud in research as a form of crime and can provide an interesting defense mechanism. To support this trend, scientists may come up with a complementary claim, perhaps best expressed by Dr. Togasatsi's (Spector's former instructor) opinion that "[t]he whole (scientific) system is based to a very high degree on trust. It may have backfired . . . but I hope we don't lose sight of the biggest issue" (McKean 1981, 28), implying that trust is maintained despite an occasional blunder in the bigger picture.

Universities were also highly successful in deflecting and reframing the rhetoric of "fraud" to issues of "ethics" and "integrity." Even the ORI's name is Office of Research Integrity. We need to be reminded that the issue is one of cheating, falsifications, fabrication, in short, conning. Would we discuss Bernard Madoff's con an issue of integrity? The result is that many researchers feel uncomfortable discussing "fraud in research" but much more comfortable discussing issues of ethics or integrity. While there definitely is a connection between these terms, they are not the same. Words carry moral weight, and the morality associated with "fraud" is not the same as that associated with "ethics" or "integrity." Our work focused on fraud in research as a specific form of deviance in an organizational context. We deliberately avoided diving too deep into complex issues of "ethics regulation" or "integrity" because we wanted to keep our focus.

Another common form of advocacy that emphasizes the integrity of science focuses attention on the fraudulent researcher as an individual. In this form of defense, the personality, background, worldview, or history of the deviant researcher are highlighted so that most of the blame falls squarely on his or her shoulders. Asimov exemplified this approach very well through the name he chose for his paper (1979): "Alas, All Human." In it he wrote, "Science itself, in the abstract, is a self-correcting, truth-seeking device. There can be mistakes and misconceptions due to incomplete or erroneous data, but the movement is always from the less true to the truer. Scientists are, however, not Science. However glorious, noble and supernaturally incorruptible Science is, scientists are, alas, all human" (133).

One instructive example involves Broad and Wade's (1982) book that was probably the first comprehensive study of a large number of cases of fraud in research. The authors argue that competition in science is so intense that it creates a strong inducement to deviate, that internal scientific controls are weak, and that fraud in science is widespread. Spokespersons for the scientific community reacted angrily. The editor of *Science 1983* (published by the American Association for the Advancement of Science) noted that nine scientists (six Nobel laureates) declined to review the book and that sociologist Robert K. Merton advised against purchasing it.[104] Biologist Norton D. Zinder's (1983) review heavily criticized the book and reminded the reader that it is encouraging to know that only thirty-four cases of fraud (reported in the book) took place over a period of two thousand years, cleverly using against the authors the low number of detected cases (per thousands of years) that the authors reported. We have, so far, only seen one contemporary review that could be considered positive.[105]

The other edge is reflected by spokespersons who express mild to alarming warning statements claiming, contrary to advocacy statements, that fraud in research is quite common. Let us examine some of these.

One glaring work was published in 1973 by Barber and colleagues. This work focuses on self-regulation of physicians when it comes to ethical issues such as informed consent of subjects. The issue of responsibility in medical research is reflected throughout the book, and its implications for medical inquiries is explored (e.g., via a survey) as benefits and risks are weighed. While Barber and his colleagues focused on research on medical experimentation involving human subjects, our data do not have such a focus. Most of our 748 cases are not about direct research on humans necessarily. Nevertheless, as this 1973 work uncovered, there were problems in controlling medical experimentation, and it also raised fascinating questions such as who gains from what? And what should be done?

In one case of fraud (Broad 1982, 482), Robert Ebert, former dean of the Harvard Medical School, was quoted in *Science* as saying that fraudulent cases in science are much more frequent than often acknowledged. Ebert's view had some previous support.

A fascinating 1976 study conducted by University of London researcher

Ian St. James-Roberts (1976) via mail questionnaires over the pages of the *New Scientist* indicated that fraud in research is quite prevalent and that there is a high likelihood that nothing will happen if fraud is suspected (469). Dr. St. James-Roberts was alarmed by the responses he received and strongly recommended that science develop "more stringent" controls.[106]

Two works in 1987 shed more light on this thorny issue. Examining the case of fraud by chemist Ronald Breslow (who was forced to withdraw three texts from a journal), chemist Pamela Zurer (1987) reflects on the controversy surrounding the question of how widespread fraud in science is. She argues that replication is not always useful, among other things because it is not considered a prestigious activity, and she suggests that stricter research guidelines be used.[107] Psychologist June Tangney's (1987) paper—based on a study of researchers in a large university—is sharper in its conclusion. It suggests that substantial rates of suspected falsification and plagiarism exist but are not accompanied by a strong desire on the part of institutions to investigate or deal with these potentially discrediting issues.

Sociologist A. C. Higgins (1983) cynically notes that researchers are extremely reluctant to admit that fudge, fraud, deceit, and deception can be easily found in the history of science and that the "bad apples" theory is simply not valid. Higgins describes vividly what he refers to as the "science game," noting that most researchers can fairly well estimate the large amount of deceit in science. He states that research has "science watchers," that is, those who construct and maintain the images of science and who are responsible for the public relations of science with other human organizations (1983, 13). Higgins documents some of the techniques used by science watchers to "cover," hide, and discount cases of fraud and deceit as they attempt to draw a positive picture in which science is portrayed as progressing gradually through the accumulation of certified knowledge by honest people.[108]

One important function that cases of fraudulent researchers fulfill is obviously in constructing and maintaining moral boundaries, in creating rigidity and stability. In line with sociologists Émile Durkheim, George Herbert Mead (1918),[109] Kai Erikson (1966)[110] and others, Ben-Yehuda (1985, 2006) has indeed pointed out that examining deviance can teach us important lessons about nondeviance. Cases such as those of Burt, Kammerer, Spector, Lysenko, Castaneda, and a host of others indeed do that. Aronson (1975, 115) demonstrated this function well in his review of Koestler's (1972) book on the Kammerer affair. Aronson wrote, "Although (the) notorious (Kammerer) scandal had reached its climax twelve years before with the suicide of the then illustrious Professor . . . I was familiar with this amazing story, for it was regularly told to biology students as an object lesson." He added that they were taught that "[i]n essence falsifying evidence is just about the worst sin that a scientist can commit since such actions threaten to destroy the very heart of the scientific system."

Exposures of cases of fraud in research were sometimes published in the prestigious journal *Science* and in the widely circulated *Discover* magazine. These revelations in print served several purposes. First, the institution of science, as represented in these journals, confesses that fraud in research can and does occur. Second, it implies that such cases of fraud are infrequent. Third, it demonstrates that these cases of fraud are detectable and that science is self-correcting. Fourth, it may give an opportunity to other researchers to express their rage and shock and offer remedies to the situation via op-eds., letters to the editor, or responses. Thus, more than anything else, the detection of fraudulent researchers can promote cohesion and integration, but it can also increase pressures to practice more rigid and stringent research.

An analysis of fraud in research necessarily focuses on unpleasant aspects of research that do not present the profession in a favorable light. One should not, however, infer from this that all scientific activity (or researchers) are suspect. We have tried to point out that the Mertonian (or the conservative) view of science is misleading and that science has a few inherent problematic areas conducive to the creation and maintenance of deviant acts. Still, the Mertonian norms can be conceptualized as an ideal, not reflecting reality in its entirety. While sociologist of science Joseph Ben-David (1977, 265) noted that the Mertonian norms "apply only to the evaluation of results to be made public," this evades the real issue, which is the existence of scientific research that is not conducted according to the strict Mertonian norms. This is similar to the distinction that author Gustav Ichheiser (1970) made between "appearances and realities," that is, between those images presented for public evaluation and what really goes on. The two need not necessarily overlap.

Despite our argument that fraud in research is probably much more common than the number of detected cases would indicate, concluding that science as such does not exist and that it is all fraudulent, manipulated, or faked would—we would like to believe—be a gross misinterpretation. When we focus on analyzing deviance, it is sometimes easy to get the false impression that the whole world may be deviant. We believe that large parts, if not an overwhelming majority, of the scientific establishment do strive to do honest research, driven mostly by curiosity and the wish to extend certified knowledge. It is our experience that communities of truthful and trustworthy research scientists do in fact exist, that there is intensive scientific activity that functions according to explicit, well-established norms, and that certified knowledge is produced and transmitted from generation to generation of researchers.

SIX | Concluding Discussion

In simple and unvarnished terms, the meaning of "misconduct" or "intentional bias" in research is "fraud." It is fraud because what those who choose to involve themselves in such indecent behavior are doing is falsifying, fabricating, or concealing data or trying to take credit for work that is not theirs. One injury caused by such activities takes place when expensive and precious resources are given to researchers or other scholars to discover or develop ideas, processes, or technologies that will be useful to humankind, and these resources are wasted—knowingly and deliberately—on fraudulent activities. In this sense, some feel this deviant practice amounts to stealing or to a con game. Nevertheless, this is not the only or even the most important injury. Another injury, much more significant, takes place when science's cultural core[1] is compromised. If the cultural core of the scientific enterprise is the search for truth, then such activities wound this soul, sometimes severely. Scientific research is based on trust and on loyalty to the scientific method, to colleagues, and to the organizations that make research possible. Fraudulent practices violate—in the most significant way—both. Thus, such "misconduct" can be mainly equated to betrayal[2] of the very essence of the soul of what science, scholarship, and research are all about.

Consequently, examining fraud in research and reactions to it gives us a golden opportunity to explore and understand some of the core elements of research and science. Moreover, such an examination does not just stop at this issue. When we began this project we were focused on creating the database, but soon it became obvious that doing that involved us, together with our research assistants, in lengthy discussions not only about conceptual and methodological issues but about more general and relevant issues as truth, deceit, the structure and function of science, the differences between disciplines, the organization of science, science organizations, relevant stakeholders, and more. This book reflects this lengthy and demanding process. The first two chapters are focused on the specific question of this book: illustrations, patterns, and exploration of the nature of fraud in research. The following chapters develop an explanation for this form of deviance in organizations. Doing that necessitated broadening the explanatory horizon into a perspective that on one hand sheds light on the organizational context and on the other hand contextualizes the issue of fraud in research within theories of deviance and control. Like the analysis of other phenomena of infractions, we suggest that understanding fraud in research requires much more than just revealing the nature, practice, type, and patterns of such behavior and then unpacking its interpretations. It requires that we examine and understand the social, organizational, and moral context in which these infractions occur.

And, indeed, our examination of fraud in research reveals some fascinating features of science itself. To begin with, and as the first chapter indicates, fraud in science and reactions to it have gone through interesting and significant transformations. Some examples include higher alerts for research fraud, the establishment of formal agencies such as the ORI, increased emphasis on research ethics, increased interest of various stakeholders, and better definitions of what fraud in research is as well as more severe sanctions (even incarceration).

Reactions to fraud can achieve some positive ends. One is a clear and unambiguous general signal stating that some behaviors are not acceptable in our society (in this case, the research community), and whoever gets involved in these practices[3] runs a high risk of experiencing unpleasant consequences. Second, by disclosing such cases to the public, research organizations can signal to the public that their operations are clean and that they do not take such behavior lightly. Doing this is not without risk. Because journalism is guided by the desire to report (some say create) sensations, public exposure may harm a research organization because consumers of the media may have the impression that "something bad" is taking place there, and not in just one case. The more such cases are publicized, the stronger such an effect will become. And, indeed, quite a few cases of fraud in research have made their way into major media outlets such as the *New York Times* and others, typically framed as "scandals"[4] in science.

In 2004, professor of history and communications Ron Robin published a book titled *Scandals and Scoundrels: Seven Cases that Shook the Academy*. It focuses on "significant instances of hoax, plagiarism, and misrepresentation in the liberal arts, with special emphasis on history and anthropology" (4).[5] This focus is framed by a conceptual conjuncture whose key words are "scandals and scoundrels." In his discussion of the meaning of the scandals he describes, Robin emphasizes the crucial role played by the mass media and the Internet.

A second risk of public exposure of cases of misconduct is that research organizations may create and impose control mechanisms that may stifle the creativity and initiative of researchers. Such a controlling mechanism was implemented in a university that established a central data bank to where university researchers were required to transfer all their data for possible inspection and monitoring.[6]

The major focus of the scientific enterprise is to conduct research, seek new knowledge, discover a truth, and find answers to puzzles. Getting involved in an enterprise like this typically requires attributes such as creativity, free time, resourcefulness, and, above all, freedom to explore various routes that may (or may not) lead to discoveries and new interpretations. This requires both trust in the scientific work (Barber 1987) and tolerance. Researchers and other scholars need to be trusted that they indeed seek new knowledge, are using proper and rigorous research methodologies,[7] and that in order to do that stakeholders need to exercise tolerance for a variety of idiosyncrasies. Émile Durkheim, one of 'sociology's totemic fathers, pointed out that the existence of crime is a normal social phenomenon and that "a society exempt from it is utterly impossible" (1938, 65–73). He added that even in a society of angels "faults which appear venial to the layman will create there the same scandal that the ordinary offense does in ordinary consciousness" (1938, 69). The reason for that, he argued, is because reactions to what is defined as crime can either mark the boundaries between right and wrong or open the doors for new inventions.[8] Consequently, one basic and strong element in the practice of science is trust in the institution of science. Contrary to some social organizations that are permeated with mistrust and suspicions (e.g., insurance), science is characterized by trust. Research fraud challenges this. For example, the U.S.-based National Cancer Institute set up a Clinical Trials Monitoring Branch, resorting to both scheduled and potentially surprise audits at study sites to keep tabs on researchers.[9] Setting up such an operation implies that prior assumptions about the trustworthiness and honesty of researchers and the organizations in which they work were eroded. This is a major change in the very core of science.[10] Tolerance of idiosyncrasies is a complementary requirement. To think that deviance in science can be completely eliminated is a delusion. What can be achieved is keeping infractions to a minimum, without submerging research in a Sisyphean maze of bureaucratic rules and structures.

The danger lurking here is clear. As awareness of various moral and ethical entrepreneurs and purists increases with each sensational case of fraud, demands to tighten existing controls, enact more stringent regulation, introduce more control mechanisms, do more paperwork, and establish specialized control units in science will probably increase. Potential moral panics (Goode and Ben-Yehuda 2009) in the media regarding fraud, publication of sensationalist reports about "dramatic" cases, and the continuous pressure to "do something" about the problem[11] may eventually pave the road for a dual process to unfold. On the one hand there will be a diminishing trust in research and researchers, and on the other hand there will be an increasing number of rules, regulations, control mechanisms,[12] and bureaucratic organizations or units whose main function will be to monitor research for potential fraud. Such organizations and units may have a vested interest in growing, expanding their jurisdiction, and developing new specialties. We have seen a similar process in the "war on drugs." In the early years of the twentieth century in the United States psychoactive drugs such as morphine, heroin, and marijuana were freely available to anyone who wished to purchase them. Slowly, gradually but relentlessly, bureaucratic structures were created in response to various demands of moral entrepreneurs to control the use of different consciousness-altering substances. This criminalization process is well-documented in the United States. For a short while, even the consumption of alcohol was very severely curbed. As this process intensified, other countries joined. These organizations—whose job was to battle production, dissemination, and use of various illicit substances—grew and developed expertise in the drug control business to a point that any attempts to change drug policies have encountered stubborn and relentless objections from them. The tactic is always the same: sensationalize, publicize dramatic and shocking cases, conceal evidence that does not support your position, and so on and so forth. A similar dynamic could develop in attempts to control fraud in research. If it does,[13] then science and the other fields in which research is conducted may lose much of their independence, have the image of their trustworthiness eroded significantly, and end up as bureaucratic and formalistic enterprises, their very souls altered in a regrettable fashion. One obvious result of such a hypothetical development will be the enormous social cost in terms of forgone innovations. One example of opening the door for establishing such controls is the Dutch reaction to the Stapel case of fraud.[14]

Of course, this concern is not a call for universities and research organizations to allow fraud. On the contrary. The car is a wonderful invention; it not only adds fun, ease, mobility, and flexibility into our lives, it can also save lives. Unfortunately, car accidents can end with death and terrible injuries. Is this in itself a reason to make cars illegal? Apparently not. Knowing fairly well the cost of having motorized transportation, almost all nations welcome the use of cars. Nevertheless, attempts are made to create safe cars[15] and rules of the roads that minimize the dangers, but not at the cost of making and driving cars a punishing experience. For example, many people like to drive cars very fast. While regulatory rules of the road tend to limit this desire, they do not require drivers to drive very slowly. If we are to keep the innovative, free ambiance in research-oriented organizations and to have researchers willing to risk introducing new ideas, sometimes against strong opposition, we must make sure that social control in science will not destroy this ambiance and stifle these researchers. And, indeed, having surveyed a large number of cases of fraud, virologist Alexander Kohn concludes that "science creativity demands flexibility." While we beg to differ with his following suggestion that "I am not sure whether a character trait of absolute honesty is a prerequisite or the driving force for choosing a scientific career" (1986, 195), it is clear that many innovations in science came from unexpected ventures into unchartered waters and serendipities. These, however, do not necessarily have to involve intentional fraud.[16]

Expressing his firm belief in science, and in a way supporting Kohn's view, physicist David Goodstein's book (2010) uses a few famous cases to demonstrate that while fraud in science exists, it can lead the perpetrators in different directions and that sometimes cutting corners can lead to some fascinating results. His work indeed shows that while sometimes fraud is exactly that—that is, wearing down the very soul of science—there are cases where the attribution of "fraud" is not clear and other times when it is not justified. Kohn and Goodstein undoubtedly do not support committing fraud in research, but their work does imply that researchers who stray from prescribed protocols of research, sometimes deliberately, may in fact pave the way for new innovations. Fraud is not involved in such cases. Thus, juxtaposing deviant scientists with deviant sciences has caught our attention, as can be seen in a previous chapter.

The above considerations can be easily translated into a scale of reactions to fraud. These reactions need to be proportional to the severity of the case. If incarceration and heavy monetary penalties are at one extreme, at the other end we may have shaming,[17] reprimanding, probation, and warnings.

In our examination of fraud in research we have highlighted several factors: the structure and culture of doing science, social control in science, deviant scientists, and the organizational context. We have collected, compiled, and used a large database to try and characterize the main features of fraud in research. This database is built on public and available information from all the sources we could find. Nonetheless, it presents some obvious biases. The main bias is probably that much information comes from the American Office of Research Integrity. With the exception of Denmark, other countries do not have such central information-gathering organizations,[18] and thus our data are biased toward the United States and toward the type of focus the ORI has. Given the huge amount of research done in the United States, that bias may have some justification.

One question that has haunted many researchers is whether the problem of fraud is so bad that it threatens the very core of the scientific enterprise. The cases we have and our analytical understanding of potential motivations to commit fraud, as well as the existence of some hot spots in the structure of science, leads us to suggest that the "dark figure" is probably quite substantial.

More specifically, we believe that Merton's characterization of the goal of the scientific enterprise as that of the extension of certified knowledge (1968b, 606) is a valid statement. As such, this enterprise can be guided by the spirit of his four idealized norms: universalism, communality, disinterestedness, and organized skepticism. Nevertheless, we have also seen how the actualization and practicing of these norms is far from simple. Then we have the academic career. In many universities, an academic career is subject to strong and varied pressures,[19] to the point where some refer to it as an "academic rat race." It is clear that the academic career presents a trajectory in which there are a few hot spots of pressure. It is no coincidence that cases of misconduct are found in these spots. It is thus no surprise that we have two salient age peaks in misconduct that generally fit two transitions that are sensitive points in an academic career. One simple indication that supports the suggestion that an academic career entails constant stress with regard to research and publications is that whenever we talk to colleagues about career patterns and fraud, we almost always hear stories about cases of fraud that never reached public exposure or were never even investigated.

If we examine some of the main characteristics that emerge from our frequency data, it becomes possible to try and structure a response to two questions posed to us by one of our colleagues. Can we develop a profile of fraud in research, and is it possible to predict which and under what conditions researchers will commit fraud? The first question is easier to answer.

To begin with, we found that fraud in research forms a unique and distinguishable pattern. Examining this pattern along the time continuum suggests that the known prevalence of fraud in research—in absolute numbers—has increased over the years.

The database we have enables us to construct some interesting profiles.

Research fraud seems to become public mostly through scientific or professional journals. It tends to take place in North America, and it peaked from the late 1980s to early 2000s. This conclusion is probably a result of the sample used in our study.[20] Processing suspected cases of fraud tends to take a few years and typically involves an individual male researcher rather than a group or a team. In some cases, actors, stakeholders, and organizations other than those where the case took place may get involved and influence the proceedings or the results. Appointing an ad-hoc committee to investigate suspicions seems to be a common practice. Frequency of cases tends to peak among researchers in two age groups: mid-thirties and mid-forties. While the academic background of those committing fraud is varied, most were university trained, and there is a slight likelihood that they attended top universities. Cases of fraud tend to take place in university settings, and most involve disciplines associated with medicine, primarily oncology and immunology. The overwhelming majority of reported cases involved research efforts that were funded by sources outside of the organization in which the research was conducted. Most cases involved falsification and fabrication of data, followed by misrepresentation and plagiarism, and they were most often disclosed by an advanced tenured academic collaborator or colleague from the same organization as the suspect. Overall, most disclosures came from those within the relevant academic or professional network.

The organizational profile is also instructive. Typically, suspicion of misconduct propelled the relevant organization into launching an investigation, and in the majority of cases these investigations confirmed the suspicions. The typical organizational punitive response was a temporary exclusion of the offender from academic activities and prohibiting culprits from submitting research proposals. Some cases went to court; of these, the overwhelming majority were found either guilty of all or of some of the charges against them. While many suspects denied the charges against them, about a quarter admitted wrongdoing or confessed to being involved in fraudulent research, and less than a quarter of these expressed apologies. Typically, following cases of fraud, organizations tended to launch changes such as developing new rules, establishing offices to deal with such cases, and retraining and educating academic staff.

Unfortunately, the methodology we used and our database are not sufficient to provide strong firsthand answers for the question: "Why do they do it? Why commit fraud?" Answering this question is not simple. The pressures on researchers to publish their research results and to innovate, coupled with the lack of effective social-control mechanisms, are obviously factors that need to be considered. Still, we believe[21] that while fraud in research is underreported, it is not, by any means, a major characteristic of science. Many researchers who experience the pressures and are aware of the relatively weak controls do not commit fraud. The key to understanding why in similar situations some individual researchers become deviants and others do not is probably in the complex, combined and interactive impact of pressures, weak external control, and weak individual internal controls. This combination will get some researchers into Matza's drift situation, and from there the continuation of the road to fraud is less problematic. Reasonable and more effective organizational control mechanisms, both within universities and by social-control agents such as funding agencies, journals, and research associations, and giving replication (where possible) a higher status can definitely reduce the scope of the phenomenon, but these measures will never eliminate it completely. There will always be those who will be tempted to take risks and cut corners, boast about nonexistent results, plagiarize other people's work, and otherwise behave in an unethical manner. Dreams of a deviance-free culture are no more than a fantasy.

Nevertheless, and partly because of cases of fraud in research, research organizations and universities developed initiatives that changed the form and ambiance of doing research. It seems that today issues of research integrity and fraud have gained more prominence and are drawing more attention and interest from all members of the "organizational field" associated with its conduct. These include universities, funding agencies, research-intensive organizations, scientific associations, and scientific journals. Moreover, the media and the public have also become part of this field because suspected acts of fraud get into the news and can thus become part of agenda-setting schemes the result of which can be changes in regulation of research. It seems that each one of these stakeholders has an interest in reviewing and refining its policies and practices in order to prevent or avoid future misconduct of any kind. The question is whether or not the increased information on cases of research fraud and organization's adaptation to this changing environment will lead to higher levels of monitoring and bureaucratization. What further directions are we expected to see in the field of research ethics regulation? In chapter 3, we have shown that organizational theory claims that social-control agents also act as moral entrepreneurs in their own self-interest. This, we argue, may further become a major obstacle to reforming research ethics regulation, as these include various personal and economic interests invested in the revised regulatory systems. The current increase in suspected cases of research fraud may be the result of the increased scale of social-control agents and their awareness and interests. This, we claim, is similar to the empirically found relationship between the increase in the visibility of police and increasing amounts of reported crime. This pattern is simply because there are more people looking for crime and trying to justify their jobs by recording it (this was found in many studies—one example is Marvell and Moody's [1996] work that found an impact of police on most crime types to be substantial).

We have debated the possible outcomes of the increased awareness of fraud in research and do not seem to have converging opinions. The president of the University of Utah, Arthur K. Smith, said that the government should not try to "place on universities an obligation to guarantee the correctness of their faculty's research conclusions," calling this "an impossible and self-defeating approach."[22] While President Smith may have gone a bit too far, the underlying dilemma is clear. Ben Yehuda claims that while some control mechanisms are needed and necessary, we need to be careful not to install too many bureaucratic controls and blocks that could create an ambiance that would choke creativity and turn university faculty members into small-scale researchers of nothing controversial or outside of accepted paradigms. This possibility has gained momentum especially since in recent years the scientific enterprise has entered a production form that can be easily referred to as the industrialization and bureaucratization of science.[23] Evaluating scientific production or impact is not a bad idea, but when it is pushed and crystallized into problematic bibliometric measures,[24] adding control mechanisms that are too strong will probably not add much good. Our acceptance of cars is also accepting accidents, drunk driving, air pollution, corrupting the environment (e.g., by building roads that destroy local fauna and flora), and perhaps a few other problems. Still, so far, most people feel that the benefits of having cars outweigh the cost. If we want creative, unexpected, and influential research centers, we need to be careful that the control mechanisms do not stifle the research. The cost of those who cheat, steal, and manipulate under the cloak of "science" is something we will probably have to pay for to create and maintain a flexible enough context and research ambiance in which genuinely creative researchers can push science to new frontiers.[25]

Oliver takes a different perspective, borrowing from new institutional theories of organizations and adopting the concept of "rationalized myths" suggested by Meyer and Rowan and the concept of "decoupling" in organizational systems. "Rationalized myths" are described by Meyer and Rowan (1977, 343) as institutional rules that are adopted by organizations in reaction to external pressures:

> Formal structures are not only creatures of their relational networks in the social organization. In modern societies, the elements of rationalized formal structure are deeply ingrained in, and reflect, widespread understandings of social reality. Many of the positions, policies, programs, and procedures of modern organizations are enforced by public opinion, by the views of important constituents, by knowledge legitimated through the educational system, by social prestige, by the laws, and by the definitions of negligence and prudence used by the courts. Such elements of formal structure are manifestations of powerful institutional rules which function as highly *rationalized myths*[26] that are binding on particular organizations.

Based on this formulation, the new policies that are adopted by universities, scientific journals, scientific associations, or funding agencies after incidents of research fraud can be characterized as a formal reaction to strong external pressures toward scientific integrity and stronger enforcement of these norms.

These myths are claimed to be rationalized and impersonal and are accepted beyond the discretion of individuals in organizations and are taken for granted as legitimate. As Meyer and Rowan state:

> In modern societies, the myths generating formal organizational structure have two key properties. First, they are rationalized and impersonal prescriptions that identify various social purposes as technical ones and specify in a rule-like way the appropriate means to pursue these technical purposes rationally. . . . Second, they are highly institutionalized and thus in some measure beyond the discretion of any individual participant or organization. They must, therefore, be taken for granted as legitimate, apart from evaluations of their impact on work outcomes." (1977: 343–44)

The concept of "decoupling" that depicts the creation and maintenance of gaps between formal policies and actual organizational practices is important in this context. The concept explains how organizations protect their formal structures from these external pressures by minimizing control of activities, inspections, and evaluations by handling coordination and adjustment informally (1977, 357).

In the context of issues of scientific misconduct, research fraud, and scientific integrity, the expectation is that all scientific institutions will react to the external pressures generated by the detection and publications of cases of research fraud. That is to say, there will be no university, research agency, scientific association, scientific journal, or funding agency that will not adopt policies, rules, and educational procedures aimed at enhancing research integrity. Yet it is our hope that these will be the new generators of norms of research, inspired by Merton. Decoupling in this case means that these science organizations will protect their researchers and the scientific work conducted within them from the imposition of close monitoring on research activities and heavy reporting demands that may restrict and hamper high-quality research.

Appendix: Coding the Database

This appendix lists the main variables and categories used in coding the cases, with the variables appearing in boldface.

Source: case source. Where the case was found? Categories:

1) ORI
2) ORI International
3) Books
4) Internet
5) NSF
6) The Danish Committees on Scientific Dishonesty
7) *Science Journal*
8) *The Scientist*

Date of event: When did the event take place? Fraud in research usually happens during a certain time period and not on a single day, so we indicated year and month (when available).

Date of disclosure: When was the misconduct first discovered?

Date of end of affair: When did the affair end? When did the last institution that dealt with the case finish handling it? (MMYYYY)

End of affair1: When did the affair end? When did the last institution that dealt with the case finish handling it? (YYYY)

Disclose1: Where the affair was first disclosed to the public? Name of one publication.

Name: Name of the suspect/offender.

Individual or team: Was the fraud committed by a single person or by a group (not necessarily a research team)? Categories:

0) Team
1) Individual

Gender: suspect/offender's gender. Categories:

0) Male
1) Female
2) Male and female (for group misconduct).

Age: Suspect/offender's age at the time he/she committed fraud.
Immigrant: Misconduct is in a country other than the subject's native country?
Country of origin: Country of origin of the suspect, in case the suspect is an immigrant.
Immigration: Year of immigration, in case the suspect is an immigrant.
Academic rank of suspect: Suspect/offender's academic rank.
Rank: Suspect/offender's academic rank. For group fraud—the senior person. Classified into categories:

1. Advanced tenured academic scientist
2. Untenured academic research scientist
3. Lower level research position
4. Part time, researcher and consultant
5. Postdocs
6. Administrative assistance
7. Research staff
8. Nurse
9. Others

Educational level: Suspect's educational level. For group fraud, it refers to the senior person (numeric). Classified into 4 categories:

1. PhD
2. MD/MA/MSc/MSW
3. BA/BSc/Graduate
4. Undergraduate

Seniority: Seniority of the suspect in the institution he or she worked at the time he or she was accused of fraud (the time he or she worked there up to the "date of disclosure," in years).
General seniority: Seniority of the suspect in the discipline (since his or her PhD).
Tenured: Does the suspect have tenure?

Highest degree: Where did the suspect/offender get his/her highest academic degree?
Institution name: Suspect/offender's institution.
Institution type: Institution type (nominal).
Instype: Institution type classified into categories:

1) University
2) University-affiliated hospital
3) Non-affiliated hospital
4) Private research institute/company
5) Public research institute

Institution: Is the suspect/offender's institution a university? Categories:

0) Other
1) University

Country: In which country or state (for the United States) did the affair take place?
Region: Geographic region. Categories:

1) North America
2) Europe
3) Asia
4) Latin America
5) Africa
6) Oceania
7) Middle East

Discipline: Discipline/area of research classified into major categories:

1) Medicine
2) Natural science, technology, and engineering
3) Social sciences and humanities

Study area: Discipline/area of research classified into specific categories:

11) Oncology
12) Ophthalmology
13) Neurology

14) Rheumatology
15) Psychiatry
16) Endocrinology
17) Pediatrics
18) Cardiology
19) Immunology
21) Biology / Life Sciences
22) Physics
23) Chemistry
24) Pharmacology
25) Geology
26) Astronomy
27) Engineering
28) Agriculture
29) Computer Science
31) Anthropology
32) Archeology
33) Psychology
34) Sociology
35) History
36) Social work
37) Economy
38) Education
39) Criminology
110) Diabetes research
111) Obstetrics
112) Hematology
113) Allergy research
114) Surgery
115) Virology
116) Ear, nose and throat
117) Clinical trial
118) Nutrition
119) Public health
120) Dentistry
121) Gynecology
122) Dermatology
123) Anesthesiology
124) Radiology
125) Urology

126) Nano-medicine
127) Sleep medicine
128) Internal Medicine
129) Orthopedics
221) Atmospheric sciences
222) Climate research
223) Physiology
331) Geography

Funding: Funding source classified into categories:

0) There is no funding source
1) Public
2) Private
3) Public and private

Funded: Is research funded? 0: No / 1: Yes
Published: Did the suspect publish the research?
Type of misconduct:
Falsification: 0: No / 1: Yes
Fabrication: 0: No / 1: Yes
Misrepresentation: 0: No / 1: Yes
Plagiarism: 0: No / 1: Yes
Unethical: Violation of ethical rules in research. 0: No / 1: Yes
False Credentials: Falsification of academic credentials. 0: No / 1: Yes
Paper Status: Falsification of publication status of papers. 0: No / 1: Yes
Theft: Theft of research resources. 0: No / 1: Yes
Ghost: Ghostwriting. 0: No / 1: Yes
Condescending: Condescending beliefs. 0: No / 1: Yes
Discovered: Whistle-blower classified into categories:

1) Collaborators
2) Colleague from same institution
3) Suspect's institution official
4) Funding agency official
5) Journal reviewer or editor
6) Newspaper / journalist
7) Victim
8) Mentor
9) Reviewer for the funding agency

10) Student
11) Historian of science
12) More recent colleague
13) Colleague from another institution
14) Colleagues from abroad
15) Anonymous
16) Advocacy group
17) The subject himself or herself
18) Research subjects
19) The state
20) Politician
21) Hackers
22) Reviewer and victim.
23) Colleague and victim.
24) Student and victim.
25) Collaborator and victim.

Institutional reactions, appeal and resolution:
Inquiry: The institution made an inquiry. 0: No / 1: Yes
Investigation: The institution made an investigation. 0: No / 1: Yes
Inst_result: Result of institution investigation classified into categories:

0) There was no investigation
1) Found misconduct
2) Found irregularities but no misconduct
3) Not enough evidence of misconduct
4) Didn't find misconduct or irregularities at all (total exoneration)

Appeal0: Was the decision appealed? 0: No / 1: Yes.
Appealres0: Appeal result. Categories:

0) Revoked the original decision
1) Upheld the original decision
2) Reached a settlement
3) Appellant withdrew the appeal
4) There was no appeal

Resolution1: How did the institution and the offender reach a resolution? Categories:

0) Unilateral sanctions
1) Reached an agreement, which does not necessarily entail admission of misconduct on the part of the suspect

Reprimand1: The institution reprimanded the offender. 0: No / 1: Yes
Control1: The institution requires control over the offender's future research. 0: No / 1: Yes
Controltime1: For how long was he or she being controlled (in years)?
Controlmonths1: For how long was he or she being controlled (in months)?
Restrictions: The institution put some restrictions on the offender's academic activity. 0: No / 1: Yes
Restrict time: For how long was his or her academic activity restricted (in years)?
Restrict months: For how long was his or her academic activity restricted (in months)?
Proposals: The offender was temporarily prohibited from submitting proposals for external funding. 0: No / 1: Yes
Prohibited time: For how long was he or she prohibited from submitting proposals for external funding (in years)?
Prohibited months: For how long was he or she prohibited from submitting proposals for external funding (in months)?
Positions: The institution removed the offender from some of his or her positions. 0: No / 1: Yes
Position time (years): For how long was he or she removed from his or her position (in years)?
Position months: For how long was he or she removed from his or her position (in months)?
Ethics1: The offender was forced by the institution to have ethics counseling or training. 0: No / 1: Yes
Economic: The institution imposed economic sanction. 0: No / 1: Yes
Economic sanctioning time (years): For how long was the economic sanction imposed (in years)?
Economic sanction time (in months): For how long was the economic sanction imposed (in months)?
Suspension: The institution temporarily suspended the subject. 0: No / 1: Yes
Suspension time: For how long was he or she suspended (in years)?
Suspension months: For how long was he or she suspended (in months)?
Fired: The institution fired the offender. 0: No / 1: Yes
Resign: The offender resigned. 0: No / 1: Yes

Retract1: The institution asked the offender or his or her coauthors to retract or correct a publication or grant application. 0: No / 1: Yes

Degree: Withdrawal or revocation of the offender's academic degree or rank. 0: No / 1: Yes

Inquiry2: The agency made an inquiry. 0: No / 1: Yes

Investigation2: Agency dealing with misconduct made an investigation. 0: No / 1: Yes

Agency Decision: Agency dealing with misconduct decision on the case (based on the institution or its own investigation). Categories:

1) Found fraud
2) Found irregularities but no fraud
3) Not enough evidence of fraud
4) Didn't find fraud or irregularities at all (total exoneration)
5) No further PHS action (for ORI cases only)
6) The institution failed to fulfill its reporting requirements (for ORI cases only)

Appeal: The subject appealed the agency decision. 0: No / 1: Yes

Appeal result: Appeal result. Categories:

0) Revoked the original decision
1) Upheld the original decision
2) Reached a settlement
3) Appellant withdrew the appeal
4) There was no appeal

Resolution: How did the agency and the offender reach a resolution? Categories:

0) Unilateral sanctions
1) Reached an agreement, which does not necessarily entail admission of fraud on the part of the suspect
2) There was no sanction

Closed by agency: When was the case closed by the agency?

Exclusion: The offender was temporarily excluded from some academic activities (advisory boards mostly). 0: No / 1: Yes

Exclusion time: For how long was he or she excluded (in years)?

Exclusion months: For how long was he or she excluded (in months)?
Exclusion2: The offender was permanently excluded from some academic activities. 0: No / 1: Yes
No Funds: The offender was temporarily debarred from federal funding. 0: No / 1: Yes
No funds time: For how long was he or she debarred from federal funding (in years)?
No funds months: For how long was he or she debarred from federal funding (in months)?
Reprimand: The offender was reprimanded by the agency. 0: No / 1: Yes
Control: The funding agency requested control over the offender's future research. 0: No / 1: Yes
Control time2: For how long was he or she controlled (in years)?
Control months2: For how long was he or she controlled (in months)?
Ethics: The offender was forced by the agency to have ethics counseling or training. 0: No / 1: Yes
Retract: The agency asked the offender or his or her coauthors to retract or correct a publication or grant application. 0: No / 1: Yes
Court: The case went to court. 0: No / 1: Yes
Court Decision: Trial results
Appeal1: The court ruling was appealed. 0: No / 1: Yes
Appealres1: Appeal result. Categories:

0) Revoked the original decision
1) Upheld the original decision
2) Reached a settlement
3) Appellant withdrew the appeal
4) There was no appeal

Effective punishment: Effective punishment in court (e.g., fine, prison). 0: No / 1: Yes
Punishment: Sort of punishment. Categories:

0) None
1) Prison
2) Fine
3) Prison and fine
4) House arrest
5) Probation

6) Prison and probation
7) Prison, probation, and fine
8) Probation and fine

Prison time: For how long was he or she sentenced to prison (in years)?
Prison month: For how long was he or he sentenced to prison (in months)?
Probation: For how long was the subject on probation (in months)?
House arrest: For how long was the subject sentenced to house arrest (in days)?
Punish amount: What was the amount (in U.S. dollars)?

1£ = 1.45$ 1€ = 1.26$ 1$ = 3.77₪

Professional society investigation: Was the case dealt with by a professional society? 0: No / 1: Yes
Professional society sanction: Was the subject sanctioned by a professional society? 0: No / 1: Yes
Professional society decision: Professional society decision (in words).
Irregularities: In those cases where the institution or the agency found irregularities but no fraud, what kinds of irregularities were found? Categories:

1) Deviation from accepted scientific practice
2) Honest error
3) Inadequate training, supervision, or knowledge
4) Sloppiness or inattention (professional negligence)

Retraction: Offender independently retracted or corrected articles. 0: No / 1: Yes
Retraction2: Institution, journal, or involved collaborators retracted or corrected articles. 0: No / 1: Yes
Whereabouts: What happened to the subject after the affair?
Admission: The subject admitted misconduct. 0: No / 1: Yes
Apology: The subject apologized. 0: No / 1: Yes
Denial: The subject publicly denied any wrongdoing. 0: No / 1: Yes
Organizational changes: Were there any organizational changes due to the event? 0: No / 1: Yes
If so nature: Organizational changes were introduced?

Notes

PREFACE

1. E.g., see http://ori.hhs.gov/images/ddblock/vol13_no2.pdf, p. 7.

CHAPTER 1

1. http://en.wikipedia.org/wiki/Peter_Medawar

2. See Hixson 1976; Weinstein 1979, 1–2; Zuckerman 1977.

3. Culliton 1974, 1155; Hixson 1976; http://en.wikipedia.org/wiki/William_Summerlin. Another interesting case of using paints is the suspicion that George W. Stow—a South African geologist and ethnologist—falsified a South African rock painting of a hunter masquerading as an ostrich (Dowson 1993; Dowson, Tobias, and Lewis-Williams 1994).

4. In a later testimony (Woolf 1981, 11), Long took the blame upon himself and did not accuse the system.

5. *Science*, March 6, 1981; Broad and Wade 1982, 89–96; http://www.sciencemag.org/cgi/reprint/sci%3B212/4491/137.pdf; see also http://www.albany.edu/~scifraud/data/sci_fraud_3426.html

6. Broad 1982; Wallis 1983; http://en.wikipedia.org/wiki/John_Darsee. According to Wikipedia, while Darsee involved himself in a clinical setting, he was not involved in research.

7. Darsee, John. 1983. "A Retraction of Two Papers on Cardiomyopathy," *New England Journal of Medicine* 308 (23) (June 9): 1419.

8. Director of the cancer agency.

9. Senator Orrin G. Hatch of Utah, the Republican chairman of the Senate Labor and Human Resources Committee.

10 *New York Times*, science section, June 2, 1981, http://www.nytimes.com/1981/06/02/science/cancer-agency-faces-questions-on-waste-and-abuses.html?&pagewanted=all

11. See http://www.nytimes.com/2000/06/11/nyregion/from-a-doctor-voices-of-patients-in-poetry.html

12. See http://en.wikipedia.org/wiki/David_Baltimore#Imanishi-Kari_case and http://en.wikipedia.org/wiki/Thereza_Imanishi-Kari

13. For a postcase review see Kevles 1998. David Baltimore's account can be

found in his "Baltimore's Travels" in the Summer 2003 issue of *Issues in Science and Technology*; see http://www.issues.org/19.4/updated/baltimore.html

14. See Hamilton and Baltimore 1991; Sarasohn 1993. On other cases see Ernhart, Scarr, and Geneson 1993; LaFollette 1992, 137–55; Rivlin 2004; Rossiter 1992; Sprague 1993; Swazey and Scher 1981.

15. See http://en.wikipedia.org/wiki/Andrew_Wakefield

16. See http://www.washingtonpost.com/news/morning-mix/wp/2015/01/23/the-disneyland-measles-outbreak-and-the-disgraced-doctor-who-whipped-up-fear-about-vaccinations/?tid=hp_mm&hpid=z3 and http://www.latimes.com/opinion/editorials/la-ed-measles-disneyland-20150116-story.html

17. On the Schön case, see Goodstein 2010, 97–102; Reich 2009; and http://en.wikipedia.org/wiki/Jan_Sch%C3%B6n

18. As we shall see later, whether peer reviews can detect fraud is a good question. It boils down to whether readers of papers have the mind-set or toolbox to ring a bell when suspicion of fraud emerges. It may be easier—perhaps—for relevant peers to detect plagiarism, but having peer reviewers act like detectives is perhaps an unreal expectation. In fact, our data indicate that only a small proportion of cases of suspected fraud emerged from reviews, and even there the majority of cases came from reviews of research proposals.

19. http://www.grg.org/HwangWooSuk1.htm. See also https://en.wikipedia.org/wiki/Hwang_Woo-suk

20. See http://en.wikipedia.org/wiki/Hwang_Woo-Suk#cite_note-1

21. See http://www.nytimes.com/2014/03/01/world/asia/scientists-new-project-rebuild-after-cloning-disgrace.html?_r=0. He was also interviewed in his lab: http://www.npr.org/sections/health-shots/2015/09/30/418642018/disgraced-scientist-clones-dogs-and-critics-question-his-intent?utm_medium=RSS&utm_campaign=researchnews

22. See http://ori.dhhs.gov/misconduct/cases/poehlman_notice.shtml; the *Boston Globe* had good coverage of the case. See http://www.boston.com/news/nation/articles/2005/03/18/researcher_admits_fraud_in_grant_data/ and Wikipedia: http://en.wikipedia.org/wiki/Eric_Poehlman. See also Dahlberg and Mahler 2006 and Interlandi 2006.

23. He was working at the Edinburgh branch of that U.S. pharmaceutical firm. http://retractionwatch.wordpress.com/2013/04/17/uk-researcher-who-faked-data-gets-three-months-in-jail/. See also http://www.bbc.co.uk/news/uk-scotland-edinburgh-east-fife-22186220#?utm_source=twitterfeed&utm_medium=twitter

24. http://en.wikipedia.org/wiki/Scott_Reuben

25. http://www.nytimes.com/2012/01/12/science/fraud-charges-for-dipak-k-das-a-university-of-connecticut-researcher.html?_r=0. See also http://chronicle.com/blogs/percolator/uconn-investigation-finds-that-health-researcher-fabricated-data/28291; http://healthland.time.com/2012/01/13/great-science-frauds/?slide=dipak-das#dipak-das; http://today.uconn.edu/2012/01/scientific-journals-

notified-following-research-misconduct-investigation/; and https://en.wikipedia.org/wiki/Dipak_K._Das

26. See Reuters, http://www.reuters.com/article/2013/08/20/us-usa-massachusetts-crimelab-idUSBRE97J0YW20130820; the *Boston Globe*, http://www.bostonglobe.com/metro/2013/02/03/chasing-renown-path-paved-with-lies/Axw3AxwmD33lRwXatSvMCL/story.html; and http://www.huffingtonpost.com/tag/annie-dookhan

27. http://www.slate.com/articles/news_and_politics/crime/2015/10/massachusetts_crime_lab_scandal_worsens_dookhan_and_farak.html

28. E.g., see http://www.sciencemag.org/content/343/6177/1299.long; http://www.genengnews.com/gen-news-highlights/misconduct-found-in-acid-bath-stem-cell-study/81249691/; http://www.sciencemag.org/content/344/6189/1215.full.pdf; http://emboj.embopress.org/content/33/12/1285?etoc. See also http://www.washingtonpost.com/news/morning-mix/wp/2014/07/03/how-japans-most-promising-young-stem-cell-scientist-duped-the-scientific-journal-nature-and-destroyed-her-career/?hpid=z4

29. http://www.reuters.com/article/2014/08/05/us-japan-scientist-idUSKBN0G50UI20140805

30. https://www.insidehighered.com/quicktakes/2014/10/13/after-plagiarism-probe-college-strips-senators-degree. See also http://hosted.ap.org/dynamic/stories/U/US_SENATE_MONTANA_PLAGIARISM?SITE=AP&SECTION=HOME&TEMPLATE=DEFAULT and http://www.nytimes.com/2014/08/08/us/politics/john-walsh-drops-campaign-under-pressure-from-democrats.html?_r=0

31. For more on such European cases see Weber-Wulff 2014.

32. See http://takingnote.blogs.nytimes.com/2015/05/21/two-same-sex-marriage-studies-two-debunkings/?emc=eta1&_r=0 and http://www.nytimes.com/2014/12/12/health/gay-marriage-canvassing-study-science.html?_r=0. See also http://retractionwatch.com/2015/05/20/author-retracts-study-of-changing-minds-on-same-sex-marriage-after-colleague-admits-data-were-faked/ and http://blogs.discovermagazine.com/neuroskeptic/2015/06/01/problem-michael-lacours-rebuttal/#.VZqUMvmqqkq

33. See http://www.dailymail.co.uk/news/article-3146788/Ex-Iowa-State-University-scientist-gets-four-years-prison-altering-research-HIV-vaccine-mixing-RABBIT-blood-human-blood.html; http://www.nature.com/news/us-vaccine-researcher-sentenced-to-prison-for-fraud-1.17660; http://www.theguardian.com/society/2015/jul/01/iowa-scientist-hiv-breakthrough-dong-pyou-han; and http://www.washingtonpost.com/news/to-your-health/wp/2015/07/01/researcher-who-spiked-rabbit-blood-to-fake-hiv-vaccine-results-slapped-with-rare-prison-sentence/

34. See http://chronicle.com/blogs/ticker/u-of-oregon-graduate-student-admits-to-falsifying-data-for-4-papers/102573 and http://retractionwatch.com/2015/07/27/oregon-grad-student-admits-to-faking-data-in-four-neuroscience-papers/

35. http://chronicle.com/article/Alice-Goffmans-Implausible-/232491/?key=G2gmKlBiNCtAM31hOD0RZTgEOiBpZEMqZHcbP3RzbltTEg; Lewis-Kraus 2016 and http://newramblerreview.com/book-reviews/law/ethics-on-the-run; see also http://nymag.com/scienceofus/2015/06/i-fact-checked-alice-goffman-with-her-subjects.html; http://www.slate.com/articles/news_and_politics/crime/2015/06/alice_goffman_s_on_the_run_is_the_sociologist_to_blame_for_the_inconsistencies.html; and the anonymous http://pastebin.com/raw/BzN4t0VU

36. In fact this bogus claim was a main ingredient in raising suspicions about his work.

37. http://news.sciencemag.org/scientific-community/2015/11/potti-found-guilty-research-misconduct; https://www.washingtonpost.com/news/morning-mix/wp/2015/11/09/scientist-falsified-data-for-cancer-research-once-described-as-holy-grail-feds-say/; http://retractionwatch.com/2015/11/07/its-official-anil-potti-faked-data-say-feds/. See also https://en.wikipedia.org/wiki/Anil_Potti; http://www.cbsnews.com/news/deception-at-duke-fraud-in-cancer-care/; and http://www.dukechronicle.com/article/2015/05/duke-lawsuit-involving-cancer-patients-linked-anil-potti-settled

38. For more on concept see chapter 4; Hyman 1980; and http://en.wikipedia.org/wiki/Pathological_science. Examples of this kind of deviance can be found in some ESP research, some UFO research, Dänikenism, astrology, psychic healing, and more.

39. See also Charles Bosk's 2003 book, which examines medical training and looks at professional and organizational responses to medical errors. Issues of making "honest errors" and errors that reflect negligence in medical practice are discussed. It is interesting to note Bosk's distinctions among types of errors, for example technical, judgmental, or normative.

40. "The goal of the project was to assess the causes of conflict between national groups, to anticipate social breakdown and provide eventual solutions. The proposal caused much controversy among social scientists, many of whom voiced concerns that such a study was in conflict with their professional ethics." http://en.wikipedia.org/wiki/Project_Camelot

41. On misconduct in science from an ethical point of view, see D'Angelo 2012. See also Braxton, Proper, and Bayer (2011) on faculty misconduct in graduate education.

42. See also Franklin 1984.

43. An interesting and relevant example for this can be found in Wilkinson and Pickett 2011, which examines issues of inequality and health. Christopher Snowdon's 2010 work criticizes their work on the grounds that their work involves massive data manipulations. See also http://spiritleveldelusion.blogspot.co.uk/

44. Murray Goulden (University of Nottingham) has published some recent works on the Piltdown hoax in the more general contexts of "missing links," science, and the media and culture (see http://nottingham.ac.uk/horizon/people/murray.goulden).

45. *Discover*, March 1984, 8; http://en.wikipedia.org/wiki/Great_Oil_Sniffer_Hoax

46. See http://www.bbc.com/news/uk-29459896 and https://en.wikipedia.org/wiki/ADE_651

47. http://www.skepdic.com/protsch.html. See also http://ori.dhhs.gov/international/activity/germany_anthro.shtml

48. Norwegian Radium Hospital.

49. http://ori.hhs.gov/content/case-summary-sudbo-jon

50. http://ori.dhhs.gov/misconduct/cases/Sudbo.shtml. See also Marris 2006 and http://en.wikipedia.org/wiki/Jon_Sudbo

51. Probably one of the most salient forms of fraud in scientific research. See Bowers 1997; Buranen and Roy 1999; LaFollette 1992; Mallon 1989; Marsh 2007.

52. http://www.asanet.org/sites/default/files/savvy/images/asa/docs/pdf/CodeofEthics.pdf

53. Brian Martin (individual correspondence) calls our attention to another form of abuse of power in research that is focused on attacks on dissident scientists; this is hardly ever called misconduct.

CHAPTER 2

1. The ORI suggests some procedures, as do some texts (e.g., Decco 2002). Nevertheless, these are not universally abiding.

2. United Kingdom Research Integrity Office—http://ukrio.org/

3. For a concise review see https://en.wikipedia.org/wiki/Mike_Sutton_(criminologist)

4. Some may claim that conference papers may lack the stringent controls used to review papers published in reviewed prestigious journals.

5. https://www.insidehighered.com/news/2016/02/16/president-kirkwood-community-college-admits-plagiarism-speech?utm_source=Inside+Higher+Ed&utm_campaign=966a9fc75e-DNU20160216&utm_medium=email&utm_term=0_1fcbc04421–966a9fc75e-197531353#.VsLurvIQ2I8.gmail

6. The term "dark figure" is usually used to refer to this phenomenon of unreported crimes. For introductory comments see http://en.wikipedia.org/wiki/Dark_figure_of_crime and http://en.wikipedia.org/wiki/Victim_study. For an integrated review see Maguire 2007.

7. See also Hood and Sparks 1970; Downes and Rock 1982, chap. 2; Bonn (1984, 49–65). For a short introduction see http://en.wikipedia.org/wiki/Dark_figure_of_crime

8. This probably means that researchers in high-prestige research organizations may experience more pressure to win grants in order to maintain the prestigious image of the organizations. In fact, our data indicate that in 508 known cases, most of the researchers suspected of research fraud occupied advanced tenured academic ranks (28.2 percent).

9. A similar situation characterizes police criminal records.

10. Some feel uncomfortable with using Wikipedia. Excuses for discrediting Wikipedia are typically focused on issues of stability, authoritativeness, and

the way sometimes entries are made and changed. But newspapers (and media), encyclopedias, and personal knowledge have their problems too (e.g., personal biases, fabrication, ideological biases). We feel that Wikipedia, like any other source, has its advantages and disadvantages, and when data are so difficult to come by, then using as many sources as possible is a wise methodological policy.

11. Unfortunately, information about ethnicity was scarce and we could not compile enough evidence to suggest a pattern.

12. We will return briefly to this issue toward the end of the chapter.

13. http://en.wikipedia.org/wiki/Jan_Hendrik_Schön

14. See http://en.wikipedia.org/wiki/Jan_Hendrik_Sch%C3%B6n. See also http://gradstudies.engineering.utoronto.ca/files/2012/09/Schoen_Summary.pdf

15. http://scitation.aip.org/journals/doc/PHTOAD-ft/vol_55/iss_11/15_1.shtml

16. Dennis Overbye, "After Two Scandals, Physics Group Expands Ethics Guidelines," *New York Times*, November 19, 2002.

17. Kenneth Chang. "Researcher Loses Ph.D. Over Discredited Papers," *New York Times*, June 15, 2004.

18. All these universities are highly ranked in various university ranking systems. For example, the Times Higher Education World University Rankings 2015–2016 ranks Harvard number 6; UCLA number 16; Northwestern number 25, and Pittsburgh number 79 (see https://www.timeshighereducation.com/world-university-rankings/2016/world-ranking#!/page/0/length/25/sort_by/rank_label/sort_order/asc/cols/rank_only). Academic Ranking of World Universities (http://www.shanghairanking.com/ARWU2016.html) ranks 500 universities. Harvard is no. 1, UCLA no. 12, Northwestern no. 26, and Pittsburgh no. 70.

19. Bela Gipp, *Citation-based Plagiarism Detection: Detecting Disguised and Cross-language Plagiarism using Citation Pattern Analysis*, Springer Fachmedien Wiesbaden 10 (2014). See also Carroll 2013 on the prevention of plagiarism.

20. Some cases included all three accusations and thus were coded as a "yes" on all three categories.

21. Obviously it could be that more cases were funded by research funding agencies, but we could not find evidence for this.

22. For twenty-four cases in our database we have information on both collaborators from another institution who detected the misconduct and the rank of the suspected scientist.

23. http://www.cbsnews.com/stories/2008/11/13/politics/uwire/main4597816.shtml; "Ex-U. Iowa Researcher Falsified Data in a 2005 Study," CBS News, November 12, 2008.

24. Katherine Bagley, "Immunologist Faked Data," *The Scientist*, December 1, 2009 (http://www.the-scientist.com/blog/display/56192/).

25. Philip J. Hilts, "A University and 4 Scientists Must Pay for Pilfered Work," *New York Times*, May 19, 1995.

26. Ran Reznik, *Ha'aretz*, February 18, 2005, and April 8, 2005. http://www.ynet.co.il/articles/0,7340,L-3092906,00.html

27. http://www.haaretz.co.il/hasite/pages/ShArtPE.jhtml?itemNo=542411&contrassID=2&subContrassID=1&sbSubContrassID=0

28. Philip J. Hilts, "2 Universities to Pay U.S. $1.6 Million in Research Fraud Case," *New York Times*, July 23, 1994.

29. http://grants.nih.gov/grants/guide/notice-files/not94–380.html; *Findings of Scientific Misconduct NIH Guide* 23 (45), (December 23, 1994).

30. Due to the small number of cases here and their idiosyncrasy, we do not offer any additional statistics such as average or range.

31. Stephanie Saul, "Experts Conclude Pfizer Manipulated Studies," *New York Times*, October 8, (2008), http://www.nytimes.com/2008/10/08/health/research/08drug.html?partner=rssnyt&emc=r

32. Mentioned briefly earlier. See http://ori.dhhs.gov/misconduct/cases/press_release_poehlman.shtml

33. http://retractionwatch.com/2015/08/17/64-more-papers-retracted-for-fake-reviews-this-time-from-springer-journals/

34. See http://www.manoramaonline.com/advt/week/Classics/Himalayan-Fraud.htm; http://www.ias.ac.in/jarch/currsci/70/00000221.pdf; http://www.nytimes.com/1989/04/23/us/scientist-accused-of-faking-findings.html; See also *New Scientist* 129(1755) (1991): 17; *Nature*, 1993, 366 (6456): 616.

35. Ding Yimin. "Scientific Misconduct: Beijing U. Issues First-Ever Rules," *Science* 296 (#5567) (April 2002): 448.

36. Dennis Overbye, "China Moves Toward Center of the Cosmos," *Asia-Pacific–International Herald Tribune. New York Times*. June 20, 2006. URL: http://www.nytimes.com/2006/06/20/world/asia/20iht-hawking.2013699.html?pagewanted=all

37. http://chinadigitaltimes.net/2006/05/research-fraud-rampant-in-china-robert-marquand/

38. http://ctep.cancer.gov/branches/ctmb/

39. http://ori.hhs.gov/international/activity/unworthy_degree.shtml

40. http://blogs.wsj.com/health/2009/04/10/researcher-falsified-data-in-sleep-study/

41. http://ori.hhs.gov/historical-background, January 2014.

42. See http://ori.hhs.gov/annual_reports. The last report available at the time of writing this text was that of 2011.

43. However, and unlike some of ORI's statistics, we did not examine how many cases opened in one specific year and continued to the next.

44. As mentioned above, it is possible for a research misconduct case to be classified in more than one category.

CHAPTER 3

1. We thank one of our reviewers for suggesting this direction.

2. Barber et al.'s 1973 study, which found that researchers who fall behind in academic competition are more likely to take research risks, could be taken by some as supporting both Vaughn's and Merton's ideas.

3. http://www.the-scientist.com//?articles.view/articleNo/36441/title/Editor-Quits-After-Fraud-Allegations/

4. http://www.theguardian.com/science/2012/sep/13/scientific-research-fraud-bad-practice

5. See also https://www.insidehighered.com/news/2014/08/14/almost-no-education-research-replicated-ew-article-shows

6. It is interesting that Palmer does not even use the term "misconduct" throughout the book. Instead he uses what appears to be a milder concept of "wrongdoing."

7. Referring to our limitation as human beings to having a full rational understanding of our environment.

8. For example, in *PLOS Medicine*, there is a research discussion by scientists on how policy in different countries can reduce scientific misconduct. David B. Resnik and Zubin Master, "Policies and Initiatives Aimed at Addressing Research Fraud in High-Income Countries," *PLOS Medicine* 10 (3) (2013): e1001406.

9. C. Macilwain, "Scientific Misconduct: More Cops, More Robbers?" *Cell* 149 (7):1417–19.

10. Fabrication, falsification, and plagiarizing.

11. Macilwain, "Scientific Misconduct," 1418.

12. Macilwain, "Scientific Misconduct," 1419.

13. He was eventually investigated by the General Medical Council and barred from practicing medicine.

14. Macilwain, "Scientific Misconduct," 1419.

15. http://www.apa.org/science/about/psa/2011/12/diederik-stapel.aspx

16. The blog's address is: http://retractionwatch.com/. See also: http://chronicle.com/article/Meet-Retraction-Watch-the/233373/#disqus_thread and https://en.wikipedia.org/wiki/Retraction_Watch

17. Carl Zimmer, "A Sharp Rise in Retractions Prompts Calls for Reform," *New York Times*, April 16, 2012.

18. Richard Van Noorden, "Science Publishing: The Trouble with Retractions," *Nature* 478 (2011): 26–28.

19. http://www.apa.org/pubs/journals/apl/?tab=4

20. http://www.elsevier.com/wps/find/journaldescription.cws_home/622929?generatepdf=true (September 2014).

21. This is an example of the difficulty in defining misconduct. One of our reviewers alerted us that for interdisciplinary work this rule is counterproductive. For such studies, similar papers can be published in journals directed to different disciplines. This raises an issue we do not address in the book. It is not clear sometimes whether there is really a single and clear standard of misconduct or whether research integrity is a vehicle for imposing standards from one field onto researchers in others. We find this point to show that what seems to be taken for granted as misconduct is not always the case.

22. See also: http://www.elsevier.com/conflictsofinterest

CHAPTER 4

1. Parts of this chapter and the next are based on Ben-Yehuda 1985.
2. Nevertheless, one needs to be careful here because while some such artifacts have no claim to be truthful, some may still be accepted as true, such as war or so-called historical movies. There are excellent war movies such as *The Bridge on the River Kwai* that have very little to do with what happened on the Kwai, and yet there are those who accept the movie as truthful. In fact, Rosenstone (1995, 2006) suggests that the "truth" seen in a commercial movie should be considered as no better or worse—or valid—than the historical truth.
3. Indeed, in Bhutta and Crane's article (2014) Bhutta argues that research fraud needs to be thought of as a form of crime, and Crane disagrees.
4. ORI: see https://ori.hhs.gov/
5. E.g., the British Library's review of relevant works on misconduct in science (Grayson 1995).
6. Various forms of science developed in China much earlier. Some forms of science also developed in the Mayan Empire and in ancient Egypt. The science we practice today, however, is primarily a result of European developments.
7. See also Wilson 1998.
8. For example, Hempel 1965; Hempel and Oppenheim 1948; or Bechtel 1988, to name a few.
9. Meaning that light is shifted to the red end of the spectrum. See also http://en.wikipedia.org/wiki/Red_shift for a short introduction.
10. Harrison 1980; Silk 2001; Singh 2004, 440.
11. And popularized significantly by U.S. television Channel H2.
12. For a short introduction consult http://en.wikipedia.org/wiki/Odin. See also http://en.wikipedia.org/wiki/Carl_Reichenbach and http://en.wikipedia.org/wiki/Reichenbach%27s_phenomena
13. Goran 1980; Gardner 1952. See also http://en.wikipedia.org/wiki/Wilhelm_Reich#Bion_experiments_and_T-bacilli
14. Mostly in the United States.
15. Kuhnian ideas helped start never-ending debates about whether scientific changes constitute "revolutions" or "evolutions."
16. See Ben-Yehuda 2013.
17. See, e.g., *Newsday*, October 11, 1983, and interviews on the major TV networks in the New York City area.
18. See also the *New York Magazine* review of her book: http://nymag.com/scienceofus/2015/12/when-liberals-attack-social-science.html
19. See Gordon's (1982) discussion of parapsychology as a "deviant science" and how it has attempted to escape that label.
20. See the debate in *Social Studies of Science* 11 (1981): 249–57.
21. E.g., see Hartmann 1996; Tuomela 1987.
22. This seems like a term of abuse, perhaps used more commonly by defenders of corporations to stigmatize critical research that may threaten profits.

National Law Review had a different view on junk science. See also http://www.natlawreview.com/print/article/three-ways-to-spot-junk-science or http://www.natlawreview.com/article/three-ways-to-spot-junk-science

23. See also http://en.wikipedia.org/wiki/Pathological_science

24. E.g., see http://www.velikovsky.info/I._Bernard_Cohen

25. For more on this see Mcaulay 1978.

26. This does not, however, mean easier access to publishing (Cole 1970).

27. http://en.wikipedia.org/wiki/Fortean

28. In fact, the way such programs are presented (typically without critical comments) may indeed lead observers to get the impression that these knowledge claims are valid. Moreover, the amount of literature and film documentaries done on these issues is genuinely impressive.

29. See Mcaulay 1978 and the journal *Kronos* (preceded by *Pensée*).

30. Somehow the academy felt that the word "unconventional" was better than "deviant." Sometime the term "fringe science" is invoked (see http://en.wikipedia.org/wiki/Fringe_science).

31. See, for example, Gardner 1952, 1981; Tiryakian 1974; Frazier 1981; Hanen, Osler, and Weyant 1980; Abell and Singer 1981; Nowotny and Rose 1979; Goode 2000.

32. On scandals among historians involving plagiarisms, frauds, and lies see Hoffer 2004 and Wiener 2005.

33. For a fuller description, see McKean 1981; Broad and Wade 1982, 63–73; and Kolata 1981.

34. See "A Fraud That Shook the World of Science," *New York Times*, November 1, 1981, http://www.nytimes.com/1981/11/01/magazine/a-fraud-that-shook-the-world-of-science.html

35. Nye 1980; Asimov 1979; Zuckerman 1977, 193; Gardner 1952, 745; for a short description see http://en.wikipedia.org/wiki/N_ray

36. The rise of epigenetics (see https://en.wikipedia.org/wiki/Epigenetics for a short introduction) may force us to accept some partial rehabilitation of Lamarckian ideas, although certainly not in the form that Lamarck expressed them.

37. For a simplified introduction to "Lysekoism" see http://en.wikipedia.org/wiki/Lysenkoism. As what we saw in this case was a convergence of the "politics" with the "science" one could argue—by extension—that the government was involved in this fraud as well.

38. See for example Enriquez and Gullans 2015.

39. We need to hastily add that this is not always the case. Such behavior may sometimes lead to new insights. As Mitroff's 1974 work on the commitment of scientists working on the Apollo moon project points out, confirmation bias may be common in science.

40. Kinsey's work caused quite a turmoil. See for example Allyn 1996; Bancroft 2004; Bullough 1998; Gagnon 1998; Irvine 2002; Morantz 1977; and Palmore 1952.

41. See the *New York Times*, April 26, 2013, and http://en.wikipedia.org/wiki/

Diederik_Stapel. See also http://www.nytimes.com/2011/11/03/health/research/noted-dutch-psychologist-stapel-accused-of-research-fraud.html?_r=0; http://retractionwatch.com/category/by-author/diederik-stapel/; and http://www.academia.edu/6769553/Changes_in_publication_culture_and_the_Stapel_fraud_case; on the Stapel case as indicating serious problems of self-correction in research see https://areycorneja.wordpress.com/2012/12/05/scientific-misconduct-and-the-myth-of-self-correction-in-science/. See also a translation of Stapel's own book: http://beinspired.no/wp-content/uploads/2016/03/FakingScience-20141214.pdf

42. http://en.wikipedia.org/wiki/Marc_Hauser

43. http://www.independent.co.uk/life-style/health-and-families/health-news/exposed-edward-erin-the-doctor-whose-faked-asthma-drug-test-results-proved-fatal-8662548.html

44. Large collections of tales about various hoaxes can be found in the works by MacDougall (1940), Tibballs (2006), and Blundell and Blackhall (2009). See Frazier 2009, 235–361, for a collection of hoaxes, frauds, and questionable works in science.

45. Sifakis's book (1996) even details a large number of hoaxes using a comic book style.

46. For example MacDougall 1940, 204–9; Tibballs 2006, 230–36.

47. Bartholomew and Radford 2012; Fedler 1989; Irving 2006.

48. See Ted Scheinman's 2015 AEON paper "Why Do Archaeological Fraudsters Work So Hard to Deceive Us? Because Bad Science Makes for Good Stories," http://aeon.co/magazine/society/why-do-archaeological-fraudsters-deceive-us/?utm_source=Aeon+newsletter&utm_campaign=fb83b295d7-Weekly_Newsletter_31st_July_7_30_2015&utm_medium=email&utm_term=0_411a82e59d-fb83b295d7-68709097

49. In October 2013 a new Sokal-type hoax was tried. Some journals received a nonsense paper and agreed to publish it. See Bohannon 2013.

50. For example Lindskoog 1993.

51. For example Jones 1990; Lurie 2004.

52. Innes 2005; Fritze 2009; see also Shenkman 1993.

53. Hoffer 2004; Wiener 2005.

54. https://en.wikipedia.org/wiki/Victor_Lustig

55. See, for example, Kossy 1994; Shermer 2002.

56. See Robinson 1996; *Social Research*, Fall 1996.

57. On practicing deception by the military see, for example, Barnes 1994, 23–29; Bendeck 2013; Broad 1992; Brown 1975; Charters and Tugwell 1990; Dear and Foot 1995, 283–86; Dunnigan and Nofi 1995; Kilzer 1994; Handel 1987; Howard 1990; Lloyd 1997; Mure 1980; Rankin 2008; Reit 1979; Sexton 1986; Sykes 1991; Young and Stamp 1989; Wheatley 1980.

58. Charters and Tugwell 1990; Friedman 2000.

59. Combined with social media, some may in fact feel that they are living in some sort of a posttruth period (e.g., see https://en.wikipedia.org/wiki/Post-truth_politics#References).

60. When a group of people witnesses an event, each may describe it differently. Even when they are trying to be as accurate as possible different versions about what actually took place may emerge. The question of what "really" happened may thus become a thorny issue. Courts, obviously, have to deal with this problem almost daily. Akira Kurosawa's 1950 *Rashomon* reminds us that this issue in nothing but new.

61. Department of Coins and Medals.

62. The Israel Museum devoted a whole issue of its quarterly *Mishkafaim* (Eyeglasses) to the subject of forgery in the arts and related topics of forgery in the sciences, politics, in police work, and in nature (February 1995, issue no. 23, in Hebrew).

63. E.g., see http://www.albany.edu/~scifraud/data/sci_fraud_2927.html or http://www.unz.org/Pub/SaturdayRev-1964aug01-00042. See also Salmon 2015 for an updated version, which also compares popular and scientific journalism, and http://www.spectator.co.uk/2016/10/how-many-scientific-papers-just-arent-true/ on whether scientific papers should be trusted.

64. Using this criterion of intention, one can classify types of lies and deceptions (see Barnes' 1994 reviews).

65. Ekman's definition is consistent with another popular form of deception: the hoax.

66. Harlow 1997. In the autumn of 1997, newspapers in London—the *Times*, the *Guardian*—gave readers details and offered interpretations.

67. And a substantial discussion on the question of whether it is possible to detect if a person is lying or not. See also Ben-Yehuda 2001:40–41.

68. Detecting if a person lies or not is obviously a major issue in such areas as police investigative techniques that look at alibis, results of "lie detector" tests, facial expressions and body language, as well as simple advice on how to be a successful "lie catcher" (e.g., Ekman 1992; Zebrowitz 1997).

69. To paraphrase Ekman 1992, 162. See also Jellison 1977 and works by Robert Feldman and others (e.g., 2002). Almost naturally, the topic of lies and deception elicits considerable interest in the popular press. For example, in April 1997, the British *Sunday Telegraph* and the Israeli *Ha'aretz* (on April 7, 1998, 12A) reported on Jerald M. Jellison's work-in-progress, which purportedly concluded that every person in the United States lies, on the average, two hundred times a day. Most of these lies are "small," for example, saying something like "I'm sorry to interrupt, but . . ." when the person making the apology does not really care, or blaming traffic for being late for a meeting when it is not the real reason. Jellison was quoted as stating that lies were essential because someone who does not lie would probably be perceived as subversive. The Israeli *Ma'ariv* devoted its entire April 10, 1998, supplement—more than one hundred pages—to lies, deceptions, and secrets. The competing *Yediot Aharonot* carried two long reports (July 5, 1998, 12–13, *Yours* supplement; and September 24, 1998, 10–11, *24 Hours* supplement) focused on lying. The conclusion in both was clear: we lie frequently, lies are not so bad after all, and we couldn't live in a culture in which people did not lie because it would be nightmarish. The

problem would then seem to be not whether or not we lie or how frequently, but the type, content, context, and impact of the lies we tell. Some lies are accepted and excused as a matter of daily routine; others are strictly unacceptable and their exposure is typically followed by the application of sanctions. Cohen's 1999 book supports the view that lies are prevalent and provides an overwhelming number of anecdotes and illustrations for this assertion in such diverse areas as psychology, politics, love, business, sex, art, literature, and science. See also Ben-Yehuda 2002, 206–10.

70. See *Encyclopedia Judaica*, vol. 3 (1971): 169–74; and http://en.wikipedia.org/wiki/Marranos

71. 1994, 55–64; Ben-Yehuda 1985, 168–207; Kohn 1986; Sismondo 1996.

72. 1957. See also Festinger, Riecken, and Schachter 1956.

73. See Philip Zimbardo's 2008 book on this, *The Lucifer Effect: Understanding How Good People Turn Evil* (New York: Random House). See also the 2015 movie, based on that experiment, *The Stanford Prison Experiment*.

74. This specific result was unexpected, and this research was approved by an ethics committee. Still, one may wonder. At least since the 1960s, psychologists knew about Stanley Milgram's experiments at Yale University described below (see https://en.wikipedia.org/wiki/Milgram_experiment for a short summary). These experiments showed that subjects were willing to obey authority and cause pain to what they believed were innocent others. Thus, it could perhaps be expected how the coupling of obedience to authority to a power position might end.

75. For example, abductions by UFOs or early life sexual traumas. See, for just one example, Ofshe and Watters 1994. In fact, Jeffrey Alexander and others (2004) maintain that a cultural trauma occurs "when members of a collective feel they have been subjected to a horrendous event that leaves inedible marks upon their group consciousness, marking their memories forever and changing their future identity in fundamental and irrevocable ways" (2004, 1). See also Alexander 2012.

76. See Anderson 1986; Bailey 1991, xvii.

77. For a philosophical discussion on the nature of "truth" in the context of semantic truth theories and using techniques of mathematical logic, see Yael Cohen's (1994) work.

78. E.g., see Ben-Yehuda's 1995 work on the Masada myth.

79. E.g., see Cornwell 2003; Pringle 2006; Walker 1995.

80. See also Becker's 1967 discussion on hierarchies of credibility.

81. The fictional antagonist in Orwell's *1984*—O'Brien—would have loved this statement.

82. Barnes 1994, 30–35; Robinson 1996, 8.

83. The Japanese continued negotiations with the Americans, leading them to believe that war may not have been imminent. There are numerous sources about this attack. For an introductory text only about the attack see http://en.wikipedia.org/wiki/Pearl_Harbor_attack

84. Ben-Yehuda 1990b, 181–219.

85. http://en.wikipedia.org/wiki/Geocentric_model
86. http://en.wikipedia.org/wiki/Flat_Earth
87. E.g., see http://en.wikipedia.org/wiki/Barry_Marshall
88. Ben-Yehuda 1985, 106–67.
89. See Hearnshaw 1979; http://en.wikipedia.org/wiki/Cyril_Burt
90. For some more recent texts about this case see http://www.the-scientist.com/?articles.view/articleNo/26694/title/My-Favorite-Fraud/; http://www.iib.unsam.edu.ar/php/docencia/cursos/2011/HistoryofMisconduct.pdf
91. http://en.wikipedia.org/wiki/Henri_Moissan; see also http://www.rsc.org/chemistryworld/2013/06/classic-kit-henri-moissan-furnace. Other researchers who tried to replicate Moissan's work failed.
92. For a different account see http://www.rsc.org/chemistryworld/2013/06/classic-kit-henri-moissan-furnace
93. Those interested in scientific self-delusions can find some fascinating accounts of such cases in Klotz's (1986) book. See also the fascinating case of polywater (Franks 1981); Fritze 2009; Gould 1981; Park 2000. Arens' (1979) challenge on the empirical reality of cannibalism is interesting as well.
94. See also http://en.wikipedia.org/wiki/Cyril_Burt
95. The abstract of that classic April 1951 paper states that it reports on "Results of experiments dealing with the consequences of deviation from a group standard are described. The effect of degrees of cohesiveness and relevance of the issue on the degree of rejection of a deviate is considered. The effects of these variables on communication and induction within the groups are studied in detail" (*Journal of Abnormal and Social Psychology* [2]: 190–207).
96. For more on publication biases see http://en.wikipedia.org/wiki/Publication_bias and on systemic biases see http://en.wikipedia.org/wiki/Psychology#Systemic_bias
97. For similar political practices using textbooks see http://en.wikipedia.org/wiki/Pakistani_textbooks_controversy (in Pakistan); or http://en.wikipedia.org/wiki/Japanese_textbook_controversy (in Japan); or http://en.wikipedia.org/wiki/NCERT_controversy (in India).
98. Indeed, Petroski (1985) argues that even what seem to be successful designs can end up as failures.
99. In February 1982, Vintage published another version of the book. In it Gribbin and Flagemann analyzed how, where, and why their 1974 prophesies failed. To some extent, this affair reminds one of the supposedly Mayan apocalyptic "prophecy" of the world's end in December 2012 (see http://en.wikipedia.org/wiki/Maya_end_of_the_world).
100. Shapin's public lecture at the Van Leer Institute, Jerusalem, March 3, 1982. Works on the history and sociology of science by Shapin (1982) and others (e.g. see Gould 1981) have shown how researchers have also used science to support their cultural and political views, even to the point of inventing or distorting data.
101. In fact, the NIH report on the Darsee case specifically criticized Kloner and Braunwald for not paying close enough attention to Darsee's activities and thus enabling him to perform his fraudulent acts (Wallis 1983).

102. In fact, and if at all possible, a successful development of a famous science fiction contraption—a time machine (even if just for viewing)—will eliminate the possibility of lying and hence may bring about a major social revolution (e.g., see Isaac Asimov's 1956 science fiction story "The Dead Past").

103. This theoretical approach focuses on how (and sometimes why) people create or construct meanings of, and understandings about, the world in which they live. For a short introduction to this approach see http://en.wikipedia.org/wiki/Social_constructionism and Harris 2010.

104. See Ben-Yehuda 1990b, 181–219, and 2013 on cultural cores and their individual impacts. Fishman and Lizardo (2013) also showed how macro changes affected individual cultural tastes.

CHAPTER 5

1. Description of the case is taken from http://en.wikipedia.org/wiki/Bj%C3%B8rn_Lomborg
2. http://en.wikipedia.org/wiki/Opossum
3. We did not want to use the mythical narrative that when ostriches face danger they "bury their heads in the sand" because in fact they do not (http://en.wikipedia.org/wiki/Ostrich#Attacks).
4. For short introductory comments see http://en.wikipedia.org/wiki/Fraud. It is interesting to note that some forms of what many may consider as plagiarism may not be considered as illegal. For example, supervisors publishing from their students' research, or improper understanding of proper acknowledgment practices.
5. https://ori.hhs.gov/historical-background
6. A colleague from the social sciences told us that in his research methods class, he asks students to replicate studies that were conducted using publicly available data, and they never manage to get the same results.
7. See http://retractionwatch.com/ and Montgomery and Oliver 2017. Interesting that based on data from retracted articles, Fanelli, Costas, and Larivière (2015: abstract) noted that "scientific misconduct is more likely in countries that lack research integrity policies, in countries where individual publication performance is rewarded with cash, in cultures and situations were mutual criticism is hampered, and in the earliest phases of a researcher's career."
8. Merton's norms were not based on any empirical study, and his claim that these norms were central to scientific research were more of an ideal. As could be expected from his fourth norm, many sociologists of science criticized Merton's norms in the late 1960s and early 1970s. Perhaps one more vivid illustration is Ian Mitroff's exposé of what he referred to as counternorms (1974). Nevertheless, as an ideal form it is well worth delving into these norms as well as criticize them in the context of fraud in research.
9. Excluded from this discussion are theories that view the state as a control mechanism.
10. Indeed, Bridgstock (1982) points out that there are areas in science where

social control does not operate effectively. After making a deep immersion into the history and structure of science vs. fraud, Judson (2004) feels that the increased use of the Internet in publishing science may reduce the incidence of fraud and thus act as a more effective social-control mechanism. LaFollette (1992, 203) on the other hand, feels that the new communication technologies may create new opportunities for fraud.

11. See also Wagner 2008.

12. Some may conclude that gossiping in this "Invisible College" may act as a potent social-control mechanism. But Vaidyanathan, Khalsa, and Ecklund's 2016 study of gossiping among 251 scientists pointed out that they "are not primarily worried about occasional egregious ethical violations such as fabrication, falsification, and plagiarism, and/or the processes of formal sanctions in response to these types of serious transgressions of scientific norms" (569), meaning that network gossiping is probably not an effective social-control mechanism.

13. This conversation took place in the early 1980s.

14. *Wall Street Journal*, July 13, 2014. Dreger's 2015 work only adds credibility to this warning. See also Stefano Balietti's 2016 critical review "Science Is Suffering Because of Peer Review's Big Problems," https://newrepublic.com/article/135921/science-suffering-peer-reviews-big-problems

15. This conversation took place in the early 1980s.

16. This conversation took place in the early 1980s.

17. And we cannot but mention here the science humor magazine *Journal of Irreproducible Results* (http://www.jir.com/), founded in Israel in 1955.

18. See Wolins 1962; Craig and Reese 1973; Wicherts et al. 2006. See also https://www.insidehighered.com/news/2014/08/14/almost-no-education-research-replicated-new-article-shows

19. http://www.sciencemag.org/content/349/6251/aac4716; See also https://www.insidehighered.com/news/2015/08/28/landmark-study-suggests-most-psychology-studies-dont-yield-reproducible-results; http://www.theatlantic.com/health/archive/2015/08/psychology-studies-reliability-reproducability-nosek/402466/; and in the *New York Times* http://www.nytimes.com/2015/08/28/science/many-social-science-findings-not-as-strong-as-claimed-study-says.html?_r=1. On suggestions to solve the replication crisis in psychology see http://chronicle.com/article/How-to-Fix-Psychology-s/233857. On difficulties in replicating see also Katavić 2014.

20. For more on issues of replication in psychology see Bohannon 2015; Earp and Trafimow 2015; Maxwell, Lau, and Howard 2015.

21. The acronym stands for Minnesota Multiphasic Personality Inventory.

22. http://en.wikipedia.org/wiki/Carlos_Castaneda

23. See also http://en.wikipedia.org/wiki/Richard_de_Mille

24. Arens' work was attacked, especially on the grounds that he did not take into consideration all cases and that his work was biased. One has to remember, however, that even if biased, the cases cited by Arens are instructive.

25. See also http://ori.dhhs.gov/content/closed-case-summary-2000 and http://ori.dhhs.gov/content/closed-case-summary-1994

26. http://www.boston.com/news/globe/health_science/articles/2006/11/27/oak_ridge_researchers_admit_errors/. See also Geoff Brumfiel, "Data Handling Causes Image Problem for Top Lab" *Nature* 444 (129) (November 9, 2006).

27. That was true when the quoted sentence was published. He later became the head of NIH.

28. See also http://grants.nih.gov/grants/guide/notice-files/not97-151.html and Dalton 1997.

29. We are grateful to one of the physicians involved for providing the details.

30. See http://ori.dhhs.gov/misconduct/cases/Uzelmeier.shtml; https://ecf.dcd.uscourts.gov/cgi-bin/show_public_doc?2007cv0753–24; and http://grants.nih.gov/grants/guide/notice-files/NOT-OD-07–059.html

31. See the *New York Times*, April 1, 1994, http://www.nytimes.com/1994/04/01/us/doctor-says-he-falsified-cancer-data-to-help-patients.html

32. Dr. Roger Poisson: F. Lowry, "I Have Learned My Lesson the Hard Way," *Canadian Medical Association Journal* 151 (6) (September 15, 1994): 835–37; Lawrence Altman, "Researcher Falsified Data in Breast Cancer Study," *New York Times*, March 14, 1994; Clyde H. Farnsworth "Doctor Says He Falsified Cancer Data to Help Patients," *New York Times*, April 1, 1994; John Roberts, "Fraud in Breast Cancer Trial Shakes US," *British Medical Journal* 308 (March 26, 1994): 809–10; Christine Gorman, "Breast Cancer: A Diagnosis of Deceit." *Time Magazine*, June 24, 2001. See also http://www.nejm.org/doi/full/10.1056/NEJM199405193302015

33. Case description is based on http://en.wikipedia.org/wiki/Jon_Sudb%C3%B8. See also http://ori.dhhs.gov/misconduct/cases/Sudbo.shtml; *Nature* 439 (January 19, 2006): 248–49.

34. In http://en.wikipedia.org/wiki/Jon_Sudb%C3%B8

35. Nevertheless, in one case replications raised suspicions and in the other case colleagues who examined the calculations became suspicious. Thus, *The Economist*'s suggestion may be overstated.

36. Weiner 1980; Zuckerman 1977, 100–101; http://en.wikipedia.org/wiki/Piltdown_Man

37. Interestingly enough, Merriam Webster Collegiate Dictionary of 1960 lists Piltdown Man (complete with drawing) as one of the early humanoids.

38. According to the *Chronicle of Higher Education* (March 11, 2005). See http://ori.hhs.gov/images/ddblock/vol13_no3.pdf, p. 11, and http://www.theguardian.com/science/2005/feb/19/science.sciencenews?CMP=share_btn_fb

39. For one example of a suggestion to prevent fraud in research see Gary Marcus's 2012 ideas in the *New Yorker*: http://www.newyorker.com/online/blogs/newsdesk/2012/12/cleaning-up-science.html. See also Uri Simonsohn's suggestion to post raw data at http://pages.ucsd.edu/~cmckenzie/Simonsohn2013PsychScience.pdf

40. It is interesting to note the case of cold fusion. There, a potentially most significant result failed repeated replications, became controversial, and eventually and quietly almost disappeared. Surprisingly, hardly any salient and consistent claims of misconduct or fraud were raised. For a brief description see https://en.wikipedia.org/wiki/Cold_fusion. For more see Park 2000; https://aeon.co/essays/why-do-scientists-dismiss-the-possibility-of-cold-fusion; https://www.asme.org/engineering-topics/articles/nuclear/cold-fusion-25-years-later; and http://www.forbes.com/sites/startswithabang/2016/09/23/is-cold-fusion-feasible-or-is-it-a-fraud/#5ff8ec2e1903. For an interesting side development involving a criminal murder, see http://foreignpolicy.com/2016/07/07/the-coldest-case-cold-fusion-eugene-mallove-mit-infinite-energy/

41. http://chronicle.com/article/Replication-Crisis-in/147301/

42. See discussion in chapter 2.

43. Described in an earlier chapter.

44. See, for example, Abelson 1990; Kochan and Budd 1992; Stewart and Feder 1987. LaFollette (1992) focused on peer review in the context of scientific fraud.

45. Chapter 1. See Broad 1981; Charrow 2010, 52–53; http://www.nytimes.com/1982/05/20/us/doctor-admits-filing-false-data-and-is-barred-from-us-support.html

46. Broad and Wade 1982, 89–90.

47. See, for example, the recent lawsuit involving millions of US$ against Duke University: http://www.newsobserver.com/news/local/education/article99669487.html; http://www.sciencemag.org/news/2016/09/whistleblower-sues-duke-claims-doctored-data-helped-win-200-million-grants; and http://science.sciencemag.org/content/353/6303/977

48. For short introductions see https://en.wikipedia.org/wiki/Higher_education and https://en.wikipedia.org/wiki/History_of_higher_education_in_the_United_States#Twentieth_century

49. Interestingly enough, a new form of defrauding citations has emerged as well. See Biagioli's July 2016 paper in *Nature*—http://www.nature.com/news/watch-out-for-cheats-in-citation-game-1.20246

50. For one illustration of these measurements see Harzing's Publish or Perish: http://www.harzing.com/resources/publish-or-perish

51. According to Wikipedia, "The phrase appeared in a non-academic context in the 1932 book, *Archibald Cary Coolidge: Life and Letters*, by Harold Jefferson Coolidge. In 1938, the phrase appeared in a college-related publication. According to Eugene Garfield, the expression first appeared in an academic context in Logan Wilson's book, *The Academic Man: A Study in the Sociology of a Profession*, published in 1942." See https://en.wikipedia.org/wiki/Publish_or_perish

52. Based on the Darsee case, Racker (1989) suggested that fraud committed by gifted researchers may be due to a mental unbalance.

53. In a later testimony (Woolf 1981, 11) Long specifically took the blame upon himself and did not accuse the system.

54. McCaghy (1976, 118–19) also noted that there were many forms of fraud in medicine (e.g., fraudulent devices, bogus healing).

55. See chapter 2.

56. The first four cases were told freely to one of the authors of this book. Anonymity was guaranteed to all informants.

57. Pseudonym. This conversation with first author took place in the early 1980s.

58. Pseudonym. This conversation with one of the authors of this book took place in the early 1980s.

59. Pseudonym. This conversation with one of the authors of this book took place in the early 1980s.

60. Pseudonym and pseudo-department. This conversation with one of the authors of this book took place in the early 1980s.

61. These conversations with one of the authors of this book took place in the early 1980s.

62. See http://en.wikipedia.org/wiki/Jocelyn_Bell_Burnell

63. Metz 1977, 277; for more details on this strange case, see Wade 1975. Reed's (1983) analysis indicates how problematic this affair has become to astronomers.

64. http://en.wikipedia.org/wiki/Oil_drop_experiment

65. For a detailed discussion see Goodstein 2010, 29–50.

66. Sayre's position has obviously become contested, among other reasons because the recognition of Wilkins served to acknowledge the contribution of the research group where she worked alongside that of the Cambridge duo. Another criticism is that Sayre overstated her claim because when the Nobel Prize committee recognized the work, Franklin was dead, and Nobel Prizes are not given posthumously. For some more information about this case and some of the controversies surrounding it, see the short review in http://en.wikipedia.org/wiki/Rosalind_Franklin

67. The last year information was available.

68. This problem may be more accentuated (and drawing more attention) in some areas of biomedical research, particularly so when corporate employees draft papers about proprietary drugs that are later published by university-affiliated researchers who had nothing to do with the paper. That is, the papers were not written by the people whose names appear on them. For more on this specific issue see Angell 2005; Goldacre 2012; Gøtzsche 2013; and Kassirer 2005. The comparison to the huge corporation Ikea is striking. The specific names of Ikea's designers of specific products appear in the catalog.

69. For more on remorse see Weisman 2014.

70. E.g., the 2006 case involving Steven E. Leadon, who was suspected of falsifying DNA samples and figures for experiments. See http://ori.dhhs.gov/misconduct/cases/Leadon.shtml; http://en.wikipedia.org/wiki/Steven_A._Leadon; and http://jnci.oxfordjournals.org/cgi/content/full/95/18/1358

71. See details in chapter 1.

72. https://en.wikipedia.org/wiki/Hwang_Woo-suk#Official_probe_by_Seoul_National_University_and_the_confirmation_of_fraud

73. Hwang was trained as a veterinarian and supposedly should not have been expected to be aware of the declaration. Nevertheless, once he began to work with human subjects, a reasonable expectation is that he should have become acquainted with what was involved in such work.

74. http://en.wikipedia.org/wiki/Hwang_Woo-Suk

75. http://ori.dhhs.gov/international/cases/September1996.shtml and Jeanne Rubner, "German University Withdraws Doctorate After Fraud Charge," *Nature* 382 (6587) (July 11, 1996): 104.

76. See Quirin Schiermeier, "German Chemist Sues Over Fraud Allegation," *Nature* 391 (422) (January 29, 1998). It is unclear what the result was.

77. Declan Butler, "Karolinska Institute Disowns Work of Cancer Researcher," *Nature* 388 (816) (August 28, 1997).

78. See Broad and Wade 1982; Higgins 1983.

79. See also http://en.wikipedia.org/wiki/Schliemann

80. Given Sutton's previously mentioned work, Darwin may have to be added to this list.

81. See also Merton's 1984 rebuttal.

82. Inspired by symbolic interaction, the totemic father of labeling theory, Howard Becker (1963, 9), characterized it as follows: "*social groups create deviance by making the rules whose infraction constitutes deviance*, and by applying those rules to particular people and labeling them as outsiders . . . [d]eviance is *not* a quality of the act the person commits, but rather a consequence of the application by others of rules and sanctions to an 'offender.' The deviant is one to whom that label has successfully been applied; deviant behavior is behavior that people so label . . . [d]eviance is a consequence of the response of others to a person's act" (emphasis in original). For some more introductory notes to the theory, see https://en.wikipedia.org/wiki/Labeling_theory

83. Some area-specific journals may be ranked low, but still publish good quality papers.

84. Differential association theory notes that people are exposed to differential associations in their life. Some of these associations will reinforce and be sympathetic to lead a law abiding, honest and normative lifestyle. Other associations may reinforce attitudes and behavior that are favorable to unconventional and law breaking behavior. The theory states that continuous and consistent exposure to such associations will result in a person becoming a deviant. A major work by the totemic father of this theory, Edwin Sutherland, was focused on professional thieves. For some more introductory notes to the theory, see https://en.wikipedia.org/wiki/Differential_association

85. A colleague who asked to remain anonymous reported on an anonymous survey in which he was involved and that was carried out in the 1980s among undergraduate students in a U.S. university. In that survey, 75 percent admitted to "cheating" at least once (in examinations, in plagiarism, by sending another person to take an examination). This could perhaps be taken to indicate that a counter-culture of "cheating" may exist among students. Claims about increasing cheating by students are heard at the university where the two au-

thors of this book work—Hebrew University—and have led to a stiffening of penalties. This could mean that students do in fact learn how to deviate professionally and how to "get away" with it. While this finding could, perhaps, support the differential association theory, it also seems that those who cheat are not the best students, but the weaker ones who manage to get through school by resorting to deviant practices. Many known fraud cases in science, however, do not seem to have involved weak students.

86. For one interesting illustration of an attempt to use sociological concepts to explain why scientists commit fraud, see Bechtel and Pearson 1985. David J. Miller (1992) tried to suggest some personality traits that might explain why some scientists commit fraud.

87. See Ryan and Sackrey 1995.

88. See Frazier 1976 for a short survey of older theories.

89. See also Goode 2008.

90. For just one example, Spector, who fabricated data in cancer research, "accused Racker and Vogt [his superiors in the laboratory] of unethical behavior and said 'they're going to have to live with themselves'" (McKean 1981, 23).

91. Indeed Neil Garrett and his colleagues' 2016 work supports this conclusion. See http://www.nature.com/neuro/journal/vaop/ncurrent/full/nn.4426.html

92. In his 1969 book he lumped these processes under three headlines: affinity, affiliation, and signification.

93. https://www.theguardian.com/science/2016/sep/21/cut-throat-academia-leads-to-natural-selection-of-bad-science-claims-study?CMP=fb_gu

94. Up to the 1980s there were few works on fraud in science. Broad and Wade (1982) and Kilbourne and Kilbourne (1983) were among those few. Then, slowly and gradually, more and more works were published (e.g., Blakely, Poling, and Cross 1986; Kohn 1986; Woolf 1988; Robin 2004) with the ORI giving relatively easy (but frequently quite partial) access to some cases of fraud in science.

95. Gupta (2013) makes a similar claim.

96. http://www.plosone.org/article/info%3Adoi%2F10.1371%2Fjournal.pone.0005738

97. http://www.nytimes.com/2012/10/02/science/study-finds-fraud-is-widespread-in-retracted-scientific-papers.html?_r=0; and http://www.nytimes.com/2012/04/17/science/rise-in-scientific-journal-retractions-prompts-calls-for-reform.html?pagewanted=all

98. Support for this can be found in the works by Jacobsen and Hals (1995); Swazey, Anderson, and Louis (1993); and Lock and Wells (1993).

99. http://www.spectator.co.uk/2016/10/how-many-scientific-papers-just-arent-true/. One of the authors of this book sent this paper to about thirty colleagues worldwide. No one expressed surprise.

100. See also Schlegel and Weisburd 1992; Simpson and Weisburd 2009; Weisburd, Waring, and Chayet 2001; Weisburd et al. 1991; and Payne 2012.

101. E.g., see http://www.nature.com/neuro/journal/vaop/ncurrent/full/nn.4426.html; and http://www.ynet.co.il/articles/0,7340,L-3906989,00.html

102. See http://new.huji.ac.il/page/17947 (in Hebrew).

103. Actually, the United States Congress had by the 1980s become concerned about the problem of fraud in science. The Committee on Science and Technology formed a subcommittee (chaired by Albert Gore, Jr.) to investigate this problem in 1981 (Fraud in Biomedical Research 1981).

104. See *Science and Government Report,* March 1, 1983, 3–5.

105. The book, perhaps, is not very good and falls short of its pretensions, especially in the general interpretations, suggestions, and some unfortunate overgeneralization. The book does, however, offer a fascinating account of fraud in research by two reporters who knew firsthand many of the people involved.

106. It is interesting that the term used for deviance in St. James-Roberts's study was "intentional bias." There is no question that the sample was biased and that a great deal of intentional bias probably occurs in some industrial research.

107. See our previous discussions on replication research.

108. An example is Aronson's (1975) attempt to discount Koestler's (1972) book on Kammerer (who was accused of blatant fraud in biology). Koestler implied that Kammerer was not a pathological deviant, but that he fell victim to professional infighting among biologists.

109. Using the example of thieves.

110. Using the case of witch hunts in 1692 Salem.

CHAPTER 6

1. Cultural core is the symbolic essence where values, value prioritizations and deep-rooted beliefs are located. This core is enveloped by symbolic-moral universes that consist of actual behaviors that reflect the core. For example, it is based on symbolic representations such as words, images, and ideas. The core influences the symbolic-moral universe, but the behaviors of which this universe consists constantly engage other universes and these negotiations can feed changes back to the core. For more on cultural cores see Ben-Yehuda 2013.

2. See Ben-Yehuda 2001.

3. And gets caught. . . . There are those who have either not been found out or who have not been adequately sanctioned.

4. On scandals, how to make them, and on the media see Adut 2008; Kipnis 2010; Lull and Hinerman 1997.

5. Pallone and Hennessy's 1995 edited book also focuses on deceptions in the social and behavioral sciences. They point out that the range of deception there moves from plagiarism to suppression of data. They discuss conditions favorable to committing frauds and indicate that sometimes definitions of fraud and "truth" in social interactions may become less than clear. Perhaps one possible and glaring example of suppression is Masson's (1984) suggestion that Freud suppressed the seduction theory.

6. Information disclosed to the second author by a researcher from that university in a 2014 interview.

7. Sometimes it is improving the methodologies that lead to useful results; not because the previous ones were fraudulent but because the new ones give more accurate results (e.g., better questions in a questionnaire; using new and more sensitive imaging technology).

8. For the different statements see his 1933, 70–110, and 1938, 65–73, respectively.

9. http://ctep.cancer.gov/branches/ctmb/; http://ctep.cancer.gov/branches/ctmb/clinicalTrials/monitoring_coop_ccop_ctsu.htm

10. While such monitoring may perhaps be better interpreted within the context of discussing science cultural cores, it nevertheless fits here as well. Such monitoring does not have to be of a draconian nature. Suffice it that if researchers are concerned that they are being watched, their research will be more conservative.

11. The case involving David Baltimore perhaps illustrates how a maelstrom can be made into a reality.

12. E.g., see Jaffer and Cameron 2006.

13. In fact, the September 15, 2014, issue of the *New Scientist* published a long interview with the former editor of the *British Medical Journal*, Richard Smith, in which he advocated the criminalization of research misconduct. See http://www.newscientist.com/article/mg22329864.100-its-time-to-criminalise-serious-scientific-misconduct.html#.VBcp6ufQR7c

14. https://www.insidehighered.com/news/2016/06/23/netherlands-starts-major-campaign-against-research-misconduct?utm_source=Inside+Higher+Ed&utm_campaign=20043daa84-DNU20160623&utm_medium=email&utm_term=0_1fcbc04421–20043daa84–197531353#.V2unX8VOI3g.mailto

15. Contemporary work on self-driving cars may change dramatically the experience and reality of driving, road transportation, and safety.

16. See also Judson 2004, especially chap. 1. See also Decco 2002's suggestions, with a particular emphasis on plagiarism.

17. E.g., see Braithwaite 1989; Delaney 2008.

18. As much as we are aware of.

19. Savan 1988; Thelen and Di Lorenzo 1992.

20. The nature of the data we have obviously requires caution. That so many cases were reported in the United States should perhaps not be a surprise, given the high volume of research done there (as compared to other countries) and that the ORI (whose cases have primarily to do with federally funded research) exists. Calculating the number of U.S.-reported cases per scientist or per scientific endeavor for a comparative perspective is thus too problematic. Nevertheless, interpreting the meaning of these numbers needs to be supplemented by the pressure points discussed in chapter 5.

21. And it is a belief grounded in some of the relevant surveys we reviewed earlier, on personal experiences with colleagues, and on a firm belief in the scientific enterprise.

22. "Universities to Pay U.S. $1.6 Million in Research Fraud Case," *New York Times*, July 23, 1994.

23. As noted earlier, this process is composed of using bibliometric and other measurements to evaluate a researcher's reputation, standing, productivity, and impact. Attractivity to grants in terms of the number and size of competitive research grants won by a researcher as well as students' evaluations, invitations to give talks, and sometimes even media appearances have become important issues. On top of all these we now have different ratings of universities based on objective performance measurements (e.g., major prizes) or on reputation. For a critique of more aspects of this industrial process from an economist's point of view see Mirowski 2011.

24. Using statistical analysis to evaluate impacts of fields, researchers, or specific works. For a short introduction see http://en.wikipedia.org/wiki/Bibliometrics

25. For the more sociologically inclined reader, let us only add that this view is consistent with Durkheim's position on the functions of deviance.

26. Meaning that institutional rules function as formal structures that seem rational but are *unproved or false collective belief that is used to justify a social institution*. Italics added.

Bibliography

Aaronovitchi, David. 2010. *Voodoo Histories*. New York: Riverhead Books.

Abell, G. O., and B. Singer, eds. 1981. *Science and the Paranormal*. New York: Charles Scribner's Sons.

Abelson, P. H. 1990. "Mechanisms for Evaluating Scientific Information and the Role of Peer Review." *Journal of the American Society of Information Science* 41 (3): 216–22.

Adut, Ari. 2008. *On Scandal: Moral Disturbances in Society, Politics, and Art*. Cambridge: Cambridge University Press.

Alexander, Jeffrey C. 2012. *Trauma: A Social Theory*. Cambridge: Polity Press.

Alexander, Jeffrey C., R. Eyerman, B. Giesen, N. J. Smelser, and P. Sztompka, eds. 2004. *Cultural Trauma and Collective Identity*. Berkeley: University of California Press.

Allyn, David. 1996. "Private Acts/Public Policy: Alfred Kinsey, the American Law Institute and the Privatization of American Sexual Morality." *Journal of American Studies* 30 (3): 405–28.

Anderson, Myrdene. 1986. "Cultural Concatenation of Deceit and Secrecy." In *Deception: Perspectives on Human and Nonhuman Deceit*, edited by Robert W. Mitchell and Nicholas S. Thompson, 323–48. Albany: State University of New York Press.

Andreoli, N., and J. Lefkowitz. 2009. "Individual and Organizational Antecedents of Misconduct in Organizations." *Journal of Business Ethics* 85 (3): 309–32.

Angell, Marcia. 2005. *The Truth about the Drug Companies: How They Deceive Us and What to Do about It*. New York: Random House.

Arens, William. 1979. *The Man-Eating Myth: Anthropology and Anthropophagy*. Oxford: Oxford University Press.

Armstrong, J. Scott. 1982. "Research on Scientific Journals: Implications for Editors and Authors." *Journal of Forecasting* 1:83–104.

Aronson, L. R. 1975. "The Case of the Midwife Toad." *Behavior Genetics* 5:115–25.

Asch, Solomon E. 1955. "Opinions and Group Pressure." *Scientific American* 193:31–35.

Ashworth, C. E. 1980. "Flying Saucers, Spoon Bending and Atlantis: A Structural Analysis of New Mythologies." *Sociological Review* 28 (2): 353–76.

Asimov, Isaac. 1979. "Alas, All Human." *Magazine of Fantasy and Science Fiction* 56 (6): 131–49.

Babbage, Charles. 1975. *Reflections on the Decline of Science in England and on Some of Its Causes*. New York: Scholarly (reprint of 1930 edition).

Bailey, Frederick George. 1991. *The Prevalence of Deceit*. Ithaca: Cornell University Press.

Bancroft, John. 2004. "Alfred C. Kinsey and the Politics of Sex Research." *Annual Review of Sex Research* 15 (1): 1–39.

Barber, Bernard. 1987. "Trust in Science." *Minerva* 25 (1): 123–34.

Barber, Bernard, John J. Lally, Julia Loughlin Makarushka, and Daniel Sullivan. 1973. *Research on Human Subjects: Problems of Social Control in Medical Experimentation*. New York: Russell Sage Foundation.

Barnes, Barry. 1977. "On the Reception of Scientific Beliefs." In *Sociology of Science: Selected Readings*, Barry Barnes, 269–91, Middlesex: Penguin Books.

Barnes, J. A. 1994. *A Pack of Lies. Towards A Sociology of Lying*. New York: Cambridge University Press.

Bartholomew, Robert E., and Benjamin Radford. 2012. *The Martians Have Landed*. London: McFarland & Company.

Bechtel, H. K., and W. Pearson. 1985. "Deviant Scientists and Scientific Deviance." *Deviant Behavior* 6 (3): 237–52.

Bechtel, William. 1988. *Philosophy of Science: An Overview for Cognitive Science*. Mahwah, NJ: Lawrence Erlbaum Associates.

Becker, George. 1984. "Pietism and Science: A Critique of Robert K. Merton's Hypothesis." *American Journal of Sociology* 89 (5): 1065–90.

Becker, Howard S. 1963. *Outsiders*. New York: The Free Press.

Becker, Howard S. 1967. "Whose Side Are We On?" *Social Problems* 14 (3): 239–47.

Ben-David, J. 1960. "Roles and Innovations in Medicine." *American Journal of Sociology* 65:557–86.

Ben-David, J. 1977. "Organization, Social Control and Cognitive Change in Science." In *Culture and Its Creators: Essays in Honor of Edward Shils*, edited by J. Ben-David and T. Clark, 244–65. Chicago: University of Chicago Press.

Bendeck, Whitney T. 2013. *"A" Force: The Origins of British Deception During the Second World War*. Annapolis, MD: Naval Institute Press.

Ben-Yehuda, Nachman. 1985. *Deviance and Moral Boundaries*. Chicago: University of Chicago Press.

Ben-Yehuda, Nachman. 1986a. "Deviance in Science: Towards the Criminology of Science." *British Journal of Criminology* 26 (1): 1–27.

Ben-Yehuda, Nachman. 1986b. "A Quest for the Beyond and Recentering the World: The Revival of the Occult and of Science Fiction." *Journal of Popular Culture* 20 (2): 1–16.

Ben-Yehuda, Nachman. 1990a. "Gathering Dark Secrets, Hidden and Dirty Information: Some Methodological Notes on Studying Political Assassinations." *Qualitative Sociology* 13 (4): 345–72.

Ben-Yehuda, Nachman. 1990b. *The Politics and Morality of Deviance: Moral Panics, Drug Abuse, Deviant Science and Reversed Stigmatization*. Albany: State University of New York Press.

Ben-Yehuda, Nachman. 1995. *The Masada Myth: Collective Memory and Mythmaking in Israel*. Madison: University of Wisconsin Press.

Ben-Yehuda, Nachman. 2001. *Betrayals and Treason: Violations of Trust and Loyalty*. Boulder: Westview Press.

Ben-Yehuda, Nachman. 2002. *Sacrificing Truth*. Amherst, NY: Humanity Books.

Ben-Yehuda, Nachman. 2006. "Contextualizing Deviance Within Social Change and Stability, Morality and Power." *Sociological Spectrum* 26 (6): 559–80.

Ben-Yehuda, Nachman. 2013. *Atrocity, Deviance, and Submarine Warfare: Norms and Practices during the World Wars*. Ann Arbor: University of Michigan Press.

Berkowitz, Leonard. 1971. "Reporting an Experiment. A Case Study in Levelling, Sharpening and Assimilation." *Journal of Experimental Social Psychology* 7: 237–43.

Bhutta, Zulfiqar, and Julian Crane. 2014. "Should Research Fraud Be a Crime?" *British Medical Journal* 349: g4532.

Biagioli, Mario. 2012. "Recycling Texts or Stealing Time? Plagiarism, Authorship, and Credit in Science." *International Journal of Cultural Property* 19:453–76.

Black, Donald. 1983. "Crime as Social Control." *American Sociological Review* 43:34–45.

Black, Donald, ed. 1984a. *Toward a General Theory of Social Control*, 2 vols. New York: Academic Press.

Black, Donald. 1984b. "Crime as Social Control", In *Toward a General Theory of Social Control*, vol. 2, 1–27. New York: Academic Press.

Blakely, Elbert, Alan Poling, and Jeffrey Cross. 1986. "Fraud, Fakery and Fudging: Behavior Analysis and Bad Science." In *Research Methods in Applied Behavior Analysis: Issues and Advances*, edited by A. Poling and R. W. F. Fuqua, 313–30. New York: Plenum Press.

Bliss, Michael. 1982. *The Discovery of Insulin*. Chicago: University of Chicago Press.

Blum, Alan F. 1970. "The Sociology of Mental Illness." In *Deviance and Respectability*, edited by Jack Dougas, 31–60. New York: Basic Books.

Blumer, Herbert. 1948. "A Sociologist Looks at the 'Kinsey Report'." *Ecology* 29 (4): 522–24.

Blundell, Nigel, and Sue Blackhall. 2009. *Great Hoaxers, Artful Fakers and Cheating Charlatans*. South Yorkshire: Pen & Sword Books.

Bohannon, John. 2013. "Who's Afraid of Peer Review?" *Science* 342 (6154) (October 4): 60–65, http://www.sciencemag.org/content/342/6154/60.full

Bohannon, John. 2015. "Many Psychology Papers Fail Replication Test." *Science* 349 (6251): 910–11.

Bonn, R. L. 1984. *Criminology*. New York: McGraw Hill.

Bosk, Charles L. 2003. *Forgive and Remember: Managing Medical Failure*, 2nd ed. Chicago: University of Chicago Press.

Bowers, Neal. 1997. *Words for the Taking: The Hunt for a Plagiarist*. New York: W. W. Norton.

Braithwaite, John. 1989. *Crime, Shame and Reintegration*. Cambridge: Cambridge University Press.

Braunwald, E. 1992. "Cardiology: The John Darsee Experience." In *Research Fraud in the Behavioral and Biomedical Sciences*, edited by David J. Miller and Michel Hersen, 55–79. New York: John Wiley & Sons.

Braxton, John M., Eve Proper, and Alan E. Bayer. 2011. *Professors Behaving Badly. Faculty Misconduct in Graduate Education*. Baltimore: Johns Hopkins University Press.

Bridgstock, M. 1982. "A Sociological Approach to Fraud in Science." *Australian and New Zealand Journal of Sociology* 18 (3): 364–83.

Brisman, Avi. 2010. "'Creative Crime' and the Phytological Analogy." *Crime Media Culture* 6 (2): 205–25.

Broad, William J. 1980a. "Would Be Academician Pirates Papers." *Science* 208 (4451) (June 27): 1438–40.

Broad, William J. 1980b. "Imbroglio at Yale (I): Emergence of a Fraud." *Science* 210 (4465) (October 3): 38–41.

Broad, William J. 1980c. "Imbroglio at Yale (II): A Top Job Lost." *Science* 210 (4467) (October 10): 171–73.

Broad, William J. 1981. "Fraud and the Structure of Science." *Science* 212 (4491) (April 10): 137–41.

Broad, William J. 1982. "Harvard Delays in Reporting Fraud." *Science* 215 (4532): 478–82.

Broad, William J. 1992. *Teller's War: The Top-Secret Story Behind the Star Wars Deception*. New York: Simon & Schuster.

Broad, William J., and N. Wade. 1982. *Betrayers of Truth*. New York: Simon & Schuster.

Brown, Anthony Cave. 1975. *Bodyguard of Lies*. New York: Quill/William Morrow.

Brush, S. G. 1974. "Should the History of Science Be Rated X?" *Science* 183:1164–72.

Brush, S. G. 1980. "The Chimeral Cat: Philosophy of Quantum Mechanics in Historical Perspective." *Social Studies of Science* 10:393–48.

Bullough, Vern L. 1998. "Alfred Kinsey and the Kinsey Report: Historical Overview and Lasting Contribution." *Journal of Sex Research* 35 (2): 127–31.

Buranen, Lise, and Alice M. Roy, eds. 1999. *Perspectives on Plagiarism and Intellectual Property in a Postmodern World*. Albany: State University of New York Press.

Campbell, Jeremy. 2001. *The Liar's Tale: A History of Falsehood*, New York: W. W. Norton.

Carlson, Stephen C. 2005. *The Gospel Hoax*. Waco, TX: Baylor University Press.

Carroll, Jude. 2013. *A Handbook for Deterring Plagiarism in Higher Education*, 2nd ed. Oxford: Oxford Centre for Staff and Learning Development (OCSLD).

Carroll, Michael. 1977. "Of Atlantis and Ancient Astronauts: A Structural Study of Two Modern Myths." *Journal of Popular Culture* 11 (3): 541–50.

Charrow, Robert P. 2010. *Law in the Laboratory: A Guide to the Ethics of Federally Funded Science Research*. Chicago: University of Chicago Press.

Charters, David A., and Maurice A. J. Tugwell, eds. 1990. *Deception Operations: Studies in the East-West Context*. London: Brassey's.

Cohen, Adir. 1999. "Lies." In *Psychology, Politics, Love, Society, Business, Sex, Personal Life, Art, Literature and Science*. Haifa: Amatzia. (Hebrew).

Cohen, Erik, and Eyal Ben-Ari. 1991. "Hard Choices: The Sociological Analysis of Value Incommensurability." *Human Studies* 16:267–97.

Cohen, Yael. 1994. *Semantic Truth Theories*. Jerusalem: The Magness Press, Hebrew University.

Cole, Steven. 1970. "Professional Standing and the Reception of Scientific Discoveries." *American Journal of Sociology* 76: 286–306.

Colin, Norman. 1984. "No Fraud Found in Swiss Study." *Science* 223 (4639): 913.

Conklin, J. E. 1977. *Illegal But Not Criminal*. Englewood Cliffs, NJ: Prentice Hall.

Cornwell, John. 2003. *Hitler's Scientists*. New York: Viking.

Craig, J. R., and S. C. Reese. 1973. "Retention of Raw Data: A Problem Revisited." *American Psychologist* 28: 723.

Crane, D. 1972. *Invisible Colleges*. Chicago: University of Chicago Press.

Cruickshank, Charles. 1979. *Deception in World War II*. Oxford: Oxford University Press.

Culliton, B. J. 1974. "The Sloan-Kettering Affair (II): An Uneasy Resolution." *Science* 184: 1154–57.

Culliton, B. J. 1983. "Coping with Fraud: The Darsee Case." *Science* 220 (4592): 31–35.

Currell, D., and D. B. Tracy. 2010. "Finding and Fixing Corporate Misconduct." *Risk Management* 57 (3): 33–34, 36.

Dalton, Rex. 1997. "Collins' Student Sanctioned Over 'Most Severe' Case of Fraud." *Nature* 388 (July 24): 313.

D'Angelo, John. 2012. *Ethics in Science: Ethical Misconduct in Scientific Research*. Boca Raton, FL: CRC Press.

Dahlberg, John E., and Christian C. Mahler. 2006. The Poehlman Case: Running Away from the Truth." *Science and Engineering Ethics* 12: 157–73.

Dear, I. C. B., and M. R. D. Foot, eds. 1995. *The Oxford Companion to the Second World War*. Oxford: Oxford University Press.

Decco, Wilfried. 2002. *Crisis on Campus: Confronting Academic Misconduct*. Cambridge, MA: MIT Press.

Delaney, Tim. 2008. *Shameful Behaviors*. Lanham, MD: University Press of America.

de Mille, Richard. 1980. *The Don Juan Papers. Further Castaneda Controversies*. Santa Barbara, CA: Ross-Erikson.

DiMaggio, Paul, and Walter W. Powell. 1983. "The Iron Cage Revisited: Collective Rationality and Institutional Isomorphism in Organizational Fields." *American Sociological Review* 48:147–60.

Dolby, R. G. A. 1979. "Reflections on Deviant Science." In *On the Margins of Science: The Social Construction of Rejected Knowledge*, edited by R. Wallis, 9–48. Staffordshire, UK: University of Keele Press.

Douglas, Jack. 1967. *The Social Meaning of Suicide*. Princeton: Princeton University Press.

Douglas, Jack. 1977. "Shame and Deceit in Creative Deviance." In *Deviance and Social Change*, edited by Edward Sagarin, 59–86. Beverly Hills, CA: Sage.

Douglas, J., and Waksler, F. 1982. *The Sociology of Deviance: An Introduction*. Boston: Little, Brown.

Downes, D., and P. Rock. 1982. *Understanding Deviance*. Oxford: Clarendon Press.

Dowson, T. A. 1993. "The Making of Blue Ostriches: Further Light on a Scientific Fraud." *South African Journal of Science* 89 (8): 360–61.

Dowson, Thomas A., Phillip V. Tobias, and J. David Lewis-Williams. 1994. "The Mystery of the Blue Ostriches: Clues to the Origin and Authorship of a Supposed Rock Painting." *African Studies* 53 (1): 3–38.

Dreger, Alice. 2015. *Galileo's Middle Finger: Heretics, Activists, and the Search for Justice in Science*. New York: Penguin Press.

Dunn, L. C. 1965. "Mendel, His Work and His Place in History." *Proceedings of the American Philosophical Society* 109 (4): 189–98.

Dunnigan, James F., and Albert A. Nofi. 1995. *Victory and Deceit: Dirty Tricks at War*. New York: William Morrow.

Durkheim, E. (1933) 1964. *The Division of Labor in Society*. New York: The Free Press.

Durkheim, E. 1938. *The Rules of Sociological Method*. New York: The Free Press.

Earp, B. D., and D. Trafimow. 2015. "Replication, Falsification, and the Crisis of Confidence in Social Psychology." *Frontiers in Psychology* 6.

Edelhertz, H. 1970. *The Nature, Impact and Prosecution of White Collar Crime*. Washington, DC: U.S. Department of Justice, Law Enforcement Assistance Administration.

Ekman, Paul. 1992. *Telling Lies: Clues to Deceit in the Marketplace, Politics, and Marriage*. New York: W. W. Norton.

Enriquez, Juan, and Steve Gullans. 2015. *Evolving Ourselves: How Unnatural Selection and Nonrandom Mutation are Changing Life on Earth*. New York: Current.

Erikson, K. T. 1966. *Wayward Puritans*. New York: Wiley.

Ernhart, C. B., S. Scarr, and D. F. Geneson. 1993. "On Being a Whistleblower: The Needleman Case." *Ethics and Behavior* 3 (1): 73–93.

Fang, F. C., and A. Casadevall. 2011. "Retracted Science and the Retraction Index." *Infection and Immunity* 79 (10): 3855–59.

Fanelli, Daniele. 2009. "How Many Scientists Fabricate and Falsify Research? A Systematic Review and Meta-analysis of Survey Data." *PloS one* 4 (5): e5738.

Fanelli, Daniele. 2013. "Why Growing Retractions Are (Mostly) a Good Sign." *PLoS Med* 10 (12): e1001563.

Fanelli, Daniele, Rodrigo Costas, and Vincent Larivière. 2015. "Misconduct Policies, Academic Culture and Career Stage, Not Gender or Pressures to Publish, Affect Scientific Integrity." *PloS one* 10 (6): e0127556.

Fedler, Fred. 1989. *Media Hoaxes*. Ames: Iowa State University Press.

Feldman, Robert, James A. Forrest, and Benjamin R. Happ. 2002. "Self-Presentation and Verbal Deception: Do Self-Presenters Lie More?" *Basic and Applied Social Psychology* 24 (2): 163–70.

Festinger, Leon, Henry W. Riecken, and Stanley Schachter. 1956. *When Prophecy Fails*. Minneapolis: University of Minnesota Press.

Festinger, L. (1957) 1985. *A Theory of Cognitive Dissonance*. Stanford, CA: Stanford University Press.

Fishman, Robert M., and Omar Lizardo. 2013. "How Macro-Historical Change Shapes Cultural Taste: Legacies of Democratization in Spain and Portugal." *American Sociological Review* 78 (2): 213–39.

Fletcher, Harvey. 1982. "My Work with Millikan on the Oil-Drop Experiment." *Physics Today* 35 (6): 43–47.

Fletcher, Ronald. 1991. *Science, Ideology, and the Media: The Cyril Burt Scandal.* New Brunswick, NJ: Transaction Publishers.

Frankel, Eugene. 1976. "Corpuscular Optics and the Wave Theory of Light: The Science and Politics of a Revolution in Physics." *Social Studies of Science* 6: 141–84.

Franklin, Allan. 1984. "Forging, Cooking, Trimming and Riding on the Bandwagon." *American Journal of Physics* 52 (9): 786–93.

Franks, Felix. 1981. *Polywater.* Cambridge, MA: MIT Press.

Fraud in Biomedical Research. 1981. Hearings before the Subcommittee on Investigations and Oversight of the Committee on Science and Technology. U.S. House of Representatives, 97th Congress, First Session, March 3, April 1, 1981, No. 11, Published by the U.S. Government Printing Office, Washington, DC.

Frazier, Charles E. 1976. *Theoretical Approaches to Deviance.* Columbus, OH: Charles E. Merrill.

Frazier, Kendrick, ed. 1981. *Paranormal Borderlands of Science.* Amherst, NY: Prometheus Books.

Frazier, Kendrick. 2009. *Science Under Siege: Defending Science, Exposing Pseudoscience.* Amherst, NY: Prometheus Books.

Friedman, Norman. 2000. *The Fifty Year War: Conflict and Strategy in the Cold War.* Annapolis, MD: Naval Institute Press.

Fritze, Ronald H. 2009. *Invented Knowledge: False History, Faked Science and Pseudo-Religions.* London: Reaktion Books.

Gagnon, John H. 1998. "Kinsey: Totem in the Land of Taboo." *Sexualities* 91 (1): 91–95.

Gardner, Martin. 1952. *Fads and Fallacies in the Name of Science.* New York: Dover Edition.

Gardner, Martin. 1981. *Science: Good, Bad and Bogus.* New York: Prometheus Books.

Garfield, E., and A. Welljams-Dorof. 1990. "The Impact of Fraudulent Research on the Scientific Literature: The Stephen E. Breuning Case." *Journal of the American Medical Association* 263 (10): 1424–26.

Garfinkel, Harold. 1967. *Studies in Ethnomethodology.* Englewood Cliffs, NJ: Prentice-Hall.

Garrett, Neil, Stephanie C Lazzaro, Dan Ariely, and Tali Sharot. 2016. "The Brain Adapts to Dishonesty." *Nature Neuroscience* 19: 1727–32.

Geertz, Clifford. 1973. "Thick Description: Toward an Interpretive Theory of Culture." In *The Interpretation of Cultures: Selected Essays,* 3–30. New York: Basic Books.

Geis, Gilbert. 1977. *White Collar Crime. Offenses in Business, Politics and the Professions.* New York: The Free Press.

Geison, Gerald L. 1995. *The Private Science of Louis Pasteur.* Princeton: Princeton University Press.

Gieryn, T. F. 1983. "Boundary-work and the Demarcation of Science from Non-

Science: Strains and Interests in Professional Ideologies of Scientists." *American Sociological Review* 48 (6):781–95.

Giesen, Bernhard. 2005. "Performing Transcendence in Politics: Sovereignty, Deviance, and the Void of Meaning." *Sociological Theory* 23 (3): 275–85.

Gliedman, J. 1982. "Scientists in Search of the Soul." *Science Digest* 90 (7): 77–79, 105.

Goldacre, Ben. 2012. *Bad Pharma: How Drug Companies Mislead Doctors and Harm Patients*. London: Fourth Estate.

Goode, Erich. 2000. *Paranormal Beliefs*. Prospect Heights, IL: Waveland Press.

Goode, Erich. 2008. *Out of Control: Assessing the General Theory of Crime*. Stanford, CA: Stanford University Press.

Goode, Erich, and Nachman Ben-Yehuda. 2009. *Moral Panics: The Social Construction of Deviance*. Chichester, West Sussex: Wiley-Blackwell.

Goodstein, David. 2010. *On Fact and Fraud: Cautionary Tales from the Front Lines of Science*. Princeton: Princeton University Press.

Gottfredson, Michael, and Travis Hirschi. 1990. *A General Theory of Crime*. Stanford, CA: Stanford University Press.

Goran, Morris. 1980. *Fact, Fraud, and Fantasy: The Occult and Pseudosciences*. Totowa, NJ: Littlefield, Adams.

Gordon, M. D. 1982. "How Socially Distinctive Is Cognitive Deviance in an Emergent Science? The Case of Parapsychology." *Social Studies of Science* 12:151–65.

Gøtzsche, Peter C. 2013. *Deadly Medicines and Organised Crime: How Big Pharma Has Corrupted Healthcare*. Buckinghamshire: Radcliffe Medical Press.

Gould, Stephen Jay. 1977. *Ever Since Darwin*. New York: W. W. Norton

Gould, Stephen Jay. 1981. *The Mismeasure of Man*. New York: W. W. Norton.

Grayson, Lesley. 1995. *Scientific Deception*. London: The British Library.

Greeley, Andrew M. 1970a. "Implication for the Sociology of Religion of Occult Behaviour in the Youth Culture." *Youth and Society* 2: 131–40.

Greeley, Andrew M. 1970b. "Superstition, Ecstasy and Tribal Consciousness." *Social Research* 37:203–11.

Green, Stuart P. 2006. *Lying, Cheating, and Stealing*. New York: Oxford University Press.

Greve, H. R., D. Palmer, and J. E. Pozner. 2010. "Organizations Gone Wild: The Causes, Processes, and Consequences of Organizational Misconduct." *Academy of Management Annals* 4 (1): 53–107.

Gupta, Ashwaria. 2013. "Fraud and Misconduct in Clinical Research: A Concern." *Perspectives in Clinical Research* 4 (2): 144–47.

Hamilton, D. P., and D. Baltimore. 1991. "Baltimore Throws in the Towel." *Science* 252 (5007): 768–70.

Hamilton, Richard F. 1996. *The Social Misconstruction of Reality: Validity and Verification in the Scholarly Community*. New Haven: Yale University Press.

Handel, Michael, ed. 1987. *Strategic and Operational Deception in the Second World War*. Portland: Cass.

Hanen, M. P., M. J. Osler, and R. G. Weyant, eds. 1980. *Science, Pseudo-Science and Society*. Waterloo, ON: Wilfrid Laurier University Press.

Hardwig, John. 1991. "The Role of Trust in Knowledge." *Journal of Philosophy* 88 (12): 693–708.
Harlow, John. 1997. "Scarred Family Reveal Flaws in 'Saint' du Pre." *Sunday Times*, September 14, 7.
Harris, Scott R. 2010. *What Is Constructionism? Navigating Its Use in Sociology*. Boulder: Lynne Rienner.
Harrison, E. R. 1980. "The Paradox of the Dark Night Sky." *Mercury* 9 (4): 83–101.
Hartmann, D. 1996, "Protoscience and Reconstruction." *Journal of General Philosophy of Science* 27:55–69.
Harvey, Bill. 1981. "Plausibility and the Evaluation of Knowledge: A Case Study of Experimental Quantum Mechanics." *Social Studies of Science* 11: 95–130.
Hearnshaw, L. 1979. *Cyril Burt: Psychologist*. Ithaca: Cornell University Press.
Heirich, Max. 1976. "Cultural Breakthroughs." *American Behavioral Scientist* 19 (6): 685–702.
Hempel, Carl G. 1965. *Aspects of Scientific Explanation*. New York: Free Press.
Hempel, Carl G., and Paul Oppenheim. 1948. "Studies in the Logic of Explanation." *Philosophy of Science* 15 (2): 135–75.
Higgins, A. C. 1983. "The Games of Science: Science Watching." In *The Dark Side of Science*, edited by B. K. Kilbourne and M. T. Kilbourne, 9–25. San Francisco: American Academy for the Advancement of Science.
Hixson, Joseph. 1976. *The Patchwork Mouse*. New York: Anchor/Doubleday.
Hoffer, Peter Charles. 2004. *Past Imperfect. Facts, Fiction, Fraud—American History from Bancroft and Parkman to Ambrose, Bellesiles, Ellis, and Goodwin*. New York: Public Affairs.
Honig, Benson, and Akanksha Bedi. 2012. "The Fox in the Hen House: A Critical Examination of Plagiarism among Members of the Academy of Management." *Academy of Management Learning & Education* 11 (1): 101–23.
Hood, R., and S. Sparks. 1970. *Key Issues in Criminology*. New York: McGraw Hill.
Hooper, Dick. 1978. "Castaneda: Trickster-Teacher. A Conversation with Richard De Mille." *Zetetic Scholar* 1 (1): 27–30.
Horowitz, I. L., ed. 1974. *The Rise and Fall of Project Camelot*. London: MIT Press.
Horowitz, Irving Louis, and Martin Leibowitz. 1968. "Social Deviance and Political Marginality: Towards a Redefinition of the Relation Between Sociology and Politics." *Social Problems* 15 (3): 280–97.
Howard, Michael. 1990. *Strategic Deception in the Second World War*. London: Pimlico.
Humphreys, Laud. 1972. *Out of the Closets: The Sociology of Homosexual Liberation*. Englewood Cliffs, NJ: Prentice-Hall.
Humphreys, Laud. 1975. *Tearoom Trade: Impersonal Sex in Public Places*, 2nd ed. New York: Aldine.
Hyman, R. 1980. "Pathological Science: Towards a Proper Diagnosis and Remedy." *Zetetic Scholar* 6:31–39.
Ichheiser, Gustav. 1970. *Appearances and Realities*. San Francisco: Jossey-Bass.

Innes, Brian. 2005. *Fakes & Forgeries*. New York: Reader's Digest.

Interlandi, Jeneen. 2006. "An Unwelcome Discovery." *New York Times*, October 22.

Irvine, Janice M. 2002. "Toward a 'Value Free' Science of Sex." In *Sexualities in History: A Reader*, edited by Kim M. Phillips and Barry Reah, 327–59. New York: Routledge.

Irving, Clifford. 2006. *The Hoax*. New York: Miramar Books.

Jacobsen, G., and A. Hals. 1995. "Medical Investigators' Views about Ethics and Fraud in Medical Research." *Journal of the Royal College of Physicians of London* 29 (5): 405–9.

Jaffer, Usman, and Alan E. P. Cameron. 2006. "Deceit and Fraud in Medical Research." *International Journal of Surgery* 4 (2): 122–26.

Jellison, Jerald M. 1977. *I'm Sorry, I Didn't Mean To, and Other Lies We Love to Tell*. New York: Chatham Square Press.

Jones, James H. 1997a. "Dr. Yes." *New Yorker*, August 25 and September 1, 99–113.

Jones, James H. 1997b. *Alfred Kinsey: A Public/Private Life*. New York: W. W. Norton.

Jones, Mark, ed. 1990. *Fake? The Art of Deception*. Berkeley: University of California Press.

Judson, Horace Freeland. 1979. *The Eighth Day of Creation*. New York: Simon and Schuster.

Judson, Horace Freeland. 2004. *The Great Betrayal: Fraud in Science*. Orlando: Harcourt.

Kassirer, Jerome P. 2005. *On the Take: How Medicine's Complicity with Big Business Can Endanger Your Health*. Oxford: Oxford University Press.

Katavić, Vedran. 2014. "Retractions of Scientific Publications: Responsibility and Accountability." *Biochemica Media* 24 (2): 217–22, http://www.ncbi.nlm.nih.gov/pmc/articles/PMC4083573/

Kearney, H. F. 1971. *Science and Change, 1500–1700*. London: Weidenfeld and Nicholson.

Keller, Evelyn Fox. 1983. *A Feeling for the Organism: The Life and Work of Barbara McClintock*. New York: W.H. Freeman.

Kevles, Daniel J. 1998. *The Baltimore Case: A Trial of Politics, Science, and Character*. New York: W. W. Norton.

Kilbourne, Brock K., and Kilbourne, Maria T., eds. 1983. *The Dark Side of Science*. San Francisco: American Association for the Advancement of Science.

Kilzer, Louis C. 1994. *Churchill's Deception*. New York: Simon & Schuster.

Kipnis, Laura. 2010. *How to Become a Scandal: Adventures in Bad Behavior*. New York: Henry Holt.

Klotz, I. M. 1986. *Diamond Dealers and Feather Merchants: Tales from the Sciences*. Boston: Birkhauser.

Knightley, P. 1975. *The First Casualty*. New York: Harcourt Brace.

Kochan, C. A., and J. M. Budd. 1992. "The Persistence of Fraud in the Literature: The Darsee Case." *Journal of the American Society for Information Science* 43 (7): 488–93.

Koestler, A. 1972. *The Case of the Midwife Toad*. New York: Random House.

Kohler, C. S., and Publications Policy Committee of the American Diabetes Association. 2012. "Updates to Policies and Procedures Related to Potential Scientific and Academic Misconduct in the Journals of the American Diabetes Association." *Diabetes* 61 (1): 38–39.

Kohn, Alexander. 1986. *False Prophets: Fraud and Error in Science and Medicine.* Oxford: Basil Blackwell.

Kolata, G. B. 1981. "Reevaluation of Cancer Data Eagerly Awaited." *Science* 214 (October 16): 316–18.

Kossy, Donna. 1994. *Kooks. A Guide to the Outer Limits of Human Belief.* Portland, OR: Feral House.

Kuhn, T. 1962. *The Structure of Scientific Revolutions*. Chicago: University of Chicago Press.

Kuhn, T. S. 1972. *The Essential Tension*. Chicago: University of Chicago Press.

Kumar, Malhar. N. 2008. "A Review of the Types of Scientific Misconduct in Biomedical Research." *Journal of Academic Ethics* 6 (3): 211–28.

LaFollette, Marcel C. 1992. *Stealing into Print. Fraud, Plagiarism, and Misconduct in Scientific Publishing*. Berkeley: University of California Press.

Langmuir, I. 1953. *Pathological Science*. Colloquium given at the Knolls Research Laboratory, December 18, edited by R. Holl (1968), General Electric Research and Development Center Report.

Leahey, E., and K. Montgomery. 2011. "The Meaning of Regulation in a Changing Academic Profession." In *The American Academic Profession: Transformation in Contemporary Higher Education*, 295–311. Baltimore: Johns Hopkins University Press.

Lewin, R. 1983. "A Naturalist of the Genome." *Science* 222:402–5.

Lewis-Kraus, Gideon. 2016. "The Trials of Alice Goffman" (also titled "The Changeling"). *New York Times Magazine*, January 17, 30–37, 56–60.

Lindskoog, Katherine. 1993. *Fakes, Frauds & Other Malarky*, Pasadena, CA: Hope Publishing House.

List, C. J. 1985. "Scientific Fraud: Social Deviance or Failure of Virtue?" *Science, Technology and Human Values* 10 (4): 27–36.

Lloyd, Mark. 1997. *The Art of Military Deception*. London: Leo Cooper.

Lock, Stephen, and Frank Wells, eds. 1993. *Fraud and Misconduct in Medical Research*. London: BMJ Publishing Group.

Lofland, John. 1969. *Deviance and Identity*. Englewood Cliffs, NJ: Prentice Hall.

Lull, James, and Stephen Hinerman, eds. 1997. *Media Scandals*. New York: Columbia University Press.

Lurie, Doron J. 2004. *Double Exposure*. Tel Aviv: Mapa Publishers (Hebrew).

Lyman, S. M., and M. B. Scott. 1970. *A Sociology of the Absurd*. New York: Appleton-Century Crofts.

MacDougall, Curtis D. 1940. *Hoaxes*. New York: Dover Publications.

Macfarlane, Bruce, Jingjing Zhang, and Annie Pun. 2012. "Academic Integrity: A Review of the Literature." *Studies in Higher Education* 39 (2): 339–58.

Macilwain, C. 2012. "Scientific Misconduct: More Cops, More Robbers?" *Cell* 149 (7): 1417–19.

Macintyre, Ben. 2010. *Operation Mincemeat*. London: Bloomsbury.

Mackintosh, N. J., ed. 1995. *Cyril Burt: Fraud or Framed?* Oxford: Oxford University Press.

MacLean T. 2002. "Reframing Organizational Misconduct." *Business and Society* 41 (2): 242–50.

Maguire, Mike. 2007. "Crime Data and Statistics." In *The Oxford Handbook of Criminology*, edited by Mike Maguire, Rod Morgan, and Robert Reiner, 241–301. Oxford: Oxford University Press.

Mallon, Thomas. 1989. *Stolen Words: Forays into the Origins and Ravages of Plagiarism*. New York: Ticknor & Fields.

Manuel, F. E. 1968. *A Portrait of Isaac Newton*. Cambridge, MA: Harvard University Press.

Marris, Emma. 2006. "Doctor Admits *Lancet* Study Is Fiction." *Nature* 439 (January 19): 248–49.

Marshall, Eliot. 1996. "Scientific Misconduct: Fraud Strikes Top Genome Lab." *Science* 274 (November 8): 908–10.

Marsh, Bill. 2007. *Plagiarism. Alchemy and Remedy in Higher Education*. Albany: State University of New York Press.

Martin, Brian. 1994. "Plagiarism: A Misplaced Emphasis." *Journal of Information Ethics* 3 (2): 36–47.

Martin, Brian. 2013. "Countering Supervisor Exploitation." *Journal of Scholarly Publishing* 45 (1): 74–86.

Martin, Brian. 2015. "On the Suppression of Vaccination Dissent." *Science & Engineering Ethics* 21 (1): 143–57.

Marvell, T. B., and C. E. Moody. 1996. "Specification Problems, Police Levels, and Crime Rates." *Criminology* 34 (4): 609–46.

Marx, Gary T. 1984. "Notes on the Discovery, Collection, and Assessment of Hidden and Dirty Data." In *Studies in the Sociology of Social Problem*, edited by Joseph W. Schneider and John I. Kitsuse, 78–113. Norwood, NJ: Ablex.

Masson, Jeffrey Moussaieff. 1984. *The Assault on Truth. Freud's Suppression of the Seduction Theory*. New York: Farrar, Straus and Giroux.

Matza, David. 1964. *Delinquency and Drift*. New York: John Wiley and Sons.

Matza, David. 1969. *Becoming Deviant*. Englewood Cliffs, NJ: Prentice Hall.

Mauskopf, Seymour H., ed. 1979. *The Reception of Unconventional Science*, AAAS Selected Symposium, 25, published by Westview Press.

Mauskopf Seymour H., and M. R. McVaugh. 1979. "The Controversy over Statistics in Parapsychology, 1934–1938." In *The Reception of Unconventional Science*, edited by S. H. Mauskopf, 105–23, AAAS Selected Symposium, 25, published by Westview Press.

Maxwell, Scott E., Michael Y. Lau, and George S. Howard. 2015. "Is Psychology Suffering from a Replication Crisis? What Does "Failure to Replicate" Really Mean?" *American Psychologist* 70 (6): 487.

McCaghy, Charles. H. 1976. *Deviant Behavior*. New York: MacMillan.

Mcaulay, Robert. 1978. "Velikovsky and the Infrastructure of Science: The Metaphysics of a Close Encounter." *Theory and Society* 6 (3): 313–42.

McKean, Kevin. 1981. "A Scandal in the Laboratory." *Discover* 2 (11): 18–23.

Mead, G. H. 1918. "The Psychology of Punitive Justice." *American Journal of Sociology* 23: 586–92.

Merton, Robert K. 1968a. *Social Theory and Social Structure*. New York: The Free Press.

Merton, Robert K. 1968b. "Science and Democratic Social Structure." In *Social Theory and Social Structure*, 604–15. New York: Free Press.

Merton, Robert K. 1973. *The Sociology of Science: Theoretical and Empirical Investigations*. Chicago: University of Chicago Press.

Merton, Robert K. 1984. "The Fallacy of the Latest Word: The Case of 'Pietism and Science.'" *American Journal of Sociology* 89 (5): 1091–21.

Metz, William D. 1977. "Astrophysics: Discovery and the Ubiquity of Black Holes." *Science* 195 (4275) (January 21): 276–77.

Meyer, John W., J. Boli, and G. M. Thomas. 1987. "Ontology and Rationalization in the Western Cultural Account." In *Institutional Structure: Constituting State, Society, and the Individual*, edited by George M. Thomas and John W. Meyer, 12–37. Beverly Hills, CA: Sage.

Meyer, John W., and Brian Rowan. 1977. "Institutionalized Organizations: Formal Structure as Myth and Ceremony." *American Journal of Sociology* 83 (2): 340–63.

Meyer, J. W., and W. R. Scott. 1983. "Centralization and the Legitimacy Problems of Local Government." In *Organizational Environments: Ritual and Rationality*. Beverly Hills, CA: Sage.

Milgram, Stanley. 1974. *Obedience to Authority: An Experimental View*. New York: Harper & Row.

Miller, David J. 1992. "Personality Factors in Scientific Fraud and Misconduct." In *Research Fraud in the Behavioral and Biomedical Sciences*, edited by David J. Miller and Michel Hersen, 125–39. New York: John Wiley & Sons.

Mirowski, Philip. 2011. *Science-Mart: Privatizing American Science*. Cambridge, MA: Harvard University Press.

Mitchell, Robert W., and Nicholas S. Thompson, eds. 1986. *Deception: Perspectives on Human and Nonhuman Deceit*. Albany: State University of New York Press.

Mitroff, Ian I. 1974. *The Subjective Side of Science: A Philosophical Inquiry into the Psychology of the Apollo Moon Scientists*. New York: Elsevier.

Montgomery, Kathleen, and Amalya L. Oliver. 2009. "Shifts in Guidelines for Ethical Scientific Conduct: How Public and Private Organizations Create and Change Norms of Research Integrity." *Social Studies of Science* 39, no. 1 (February): 137–55.

Montgomery, K., and A. L. Oliver. 2017. "Conceptualizing Fraudulent Studies as Viruses: New Models for Handling Retractions." *Minerva* 55 (1): 49–64.

Morantz, Regina Markell. 1977. "The Scientist As a Sex Crusader: Alfred C. Kinsey and American Culture." *American Quarterly* 29 (35): 563–89.

Mulkay, M. J. 1972a. "Conformity and Innovation in Science." In *The Sociology of Science*, by P. Halmos, 5–23. Keele: University of Keele Press.

Mulkay, M. J. 1972b. *The Social Process of Innovation: A Study in the Sociology of Science*. London: Macmillan.

Mulkay, M. J., and G. N. Gilbert. 1982. "Accounting for Error: How Scientists Construct Their Social World When They Account for Correct and Incorrect Belief." *Sociology* 16 (2): 166–83.

Mure, David. 1980. *Master of Deception*. London: William Kimber.

Nettler, G. 1972. *Explaining Crime*. New York: McGraw Hill.

Noll, Richard. 1994. *The Jung Cult: Origins of a Charismatic Movement*. New York: Free Press.

Noll, Richard. 1997. *The Arian Christ: The Secret Life of Carl Jung*. New York: Random House.

Normile, Dennis. 2016. "University of Tokyo to Investigate Data Manipulation Charges Against Six Prominent Research Groups." *Science*, September 20 (see http://www.sciencemag.org/news/2016/09/university-tokyo-investigate-data-manipulation-charges-against-six-prominent-research?utm_source=newsfromscience&utm_medium=facebook-text&utm_campaign=utokyo-7542%3C/a%3E%3C/div%3E%3C/).

Norris, Christopher. 1996. *Reclaiming Truth. Contribution to a Critique of Cultural Relativism*. Durham, NC: Duke University Press.

Nowotny, H., and H. Rose, eds. 1979. *Counter-Movements in the Sciences*. Boston: D. Reidel Publishing.

Nye, Mary Jo. 1980. "N-rays: An Episode in the History and Psychology of Science." *Historical Studies in the Physical Sciences* 11: 125–26.

Ofshe, Richard, and Ethan Watters. 1994. *Making Monsters. False Memories, Psychotherapy, and Sexual Hysteria*. New York: Charles Scribner's Sons.

Orne, Martin T. 1969. "Demand Characteristics and the Concept of Quasi-Controls." In *Artifacts in Behavioral Research*, edited by Robert Rosenthal and Ralph L. Rosnow, 143–79. New York: Academic Press.

Pallone, Nathaniel J., and James J. Hennessy. 1995. *Fraud and Fallible Judgment: Varieties of Deception in the Social and Behavioral Sciences*. New Brunswick, NJ: Transaction.

Palmer, D. 2012. *Normal Organizational Wrongdoing: A Critical Analysis of Theories of Misconduct in and by Organizations*. New York: Oxford University Press.

Palmore, Erdman. 1952. "Public Reactions to the Kinsey Report." *Social Forces* 31 (2): 165–72.

Park, Robert. 2000. *Voodoo Science*. Oxford: Oxford University Press.

Payne, Brian K. 2012. *White-Collar Crime: The Essentials*. Thousand Oaks, CA: Sage Publications.

Petroski, Henry. 1985. *To Engineer Is Human: The Role of Failure in Successful Design*. New York: St. Martin's Press.

Pfeifer, M. P., and G. L. Snodgrass. 1990. "The Continued Use of Retracted, Invalid Scientific Literature." *Journal of the American Medical Association* 263 (10): 1420–23.

Pinch, Trevor J. 1979. "Normal Explanations of the Paranormal: The Demarcation Problem and Fraud in Parapsychology." *Social Studies of Science* 9: 229–48.

Plazak, Dan. 2006. *A Hole in the Ground with a Liar at the Top*. Salt Lake City: University of Utah Press.

Polanyi, M. 1967. "The Growth of Science in Society." *Minerva* 5 (4): 533–45.

Pontell, H. J., Jesilow, P. D., Geis, G. 1982. "Policing Physicians: Practitioner Fraud and Abuse in Government Medical Programs." *Social Problems* 30 (1): 117–25.

Pozner, Jo-Ellen. 2008. "Stigma and Settling Up: An Integrated Approach to the Consequences of Organizational Misconduct for Organizational Elites." *Journal of Business Ethics* 80 (1): 141–50.

Pringle, Heather. 2006. *The Master Plan: Himmler's Scholars and the Holocaust.* New York: Hyperion.

Provine, W. B. 1973. "Geneticists and the Biology of Race Crossing." *Science* 1982 (November 23): 790–96.

Quinney, Richard. 1964. "The Study of White-Collar Crime: Toward a Re-orientation in Theory and Research." *Journal of Criminal Law, Criminology and Police Science* 55: 208–14.

Racker, E. 1989. "A View of Misconduct in Science." *Nature* 339 (6220): 91–93.

Rankin, Nicholas. 2008. *A Genius for Deception: How Cunning Helped the British Win Two World Wars.* New York: Oxford University Press.

Reed, G. 1983. "The Discovery of Pulsars." *Astronomy* 11 (12): 24–28.

Reich, Eugenie Samuel. 2009. *Plastic Fantastic. How the Biggest Fraud in Physics Shook the Scientific World.* New York: Palgrave Macmillan.

Reit, Seymour. 1979. *Masquerade. The Amazing Camouflage Deceptions of World War II.* London: Robert Hale.

Rensberger, Royce. 1977. "Fraud in Research Is a Rising Problem in Science." *New York Times*, January 23, 1, 44.

Revel, Jean-Francois. 1991. *The Flight from Truth: The Reign of Deceit in the Age of Information.* Translated by Curtis Cate. New York: Random House.

Rivlin, Solomon. 2004. *Scientific Misconduct and its Cover-Up.* Boca Raton, FL: Brown Walker Press.

Roberts, David. 1982. *Great Exploration Hoaxes.* San Francisco: Sierra Club Books.

Robin, Ron. 2004. *Scandals and Scoundrels: Seven Cases that Shook the Academy.* Berkeley: University of California Press.

Robinson, W. Peter. 1996. *Deceit, Delusion, and Detection.* London: Sage Publications.

Rosenhan, D. L. 1973. "On Being Sane in Insane Places." *Science* 179 (4070): 250–58.

Rosenstone, Robert A. 1995. *Visions of the Past: The Challenge of Film to our Idea of History.* Cambridge, MA: Harvard University Press.

Rosenstone, Robert A. 2006. *History on Film/Film on History.* London: Pearson Longman.

Rosenthal, R., et al. 1979. *Sensitivity to Nonverbal Communication: The PONS Test.* Baltimore: Johns Hopkins University Press.

Rossiter, E. J. R. 1992. "Reflections of a Whistleblower." *Nature*, 357 (6378): 434–36.

Roth, J. A. 1966. "Hired Hand Research." *The American Sociologist* 10: 190–96.

Ryan, Jake, and Charles Sackrey, eds. 1995. *Strangers in Paradise: Academics from the Working Class*, 2nd ed. New York: University Press of America.

Salmon, Felix. 2015. "Why You Can't Trust Journalism." *Fusion*, July 27, http://fusion.net/story/173383/science-cant-trust-journalism/

Sarasohn, Judy. 1993. *Science on Trial: The Whistle-Blower, the Accused, and the Nobel Laureate.* New York: St. Martin's Press.

Savan, Beth. 1988. *Science Under Siege: The Myth of Objectivity in Scientific Research*. Toronto: CBC Enterprises.

Sayre, Anne. 1975. *Rosalind Franklin and DNA*. New York: W. W. Norton.

Schmaus, Warren. 1983. "Fraud and the Norms of Human Science." *Science, Technology and Human Values* 8 (4): 12–22.

Schlegel, Kip, and David Weisburd. 1992. *White-Collar Crime Reconsidered*. Boston: Northeastern University Press.

Schlussel, Y. R. 1983. "Structural Determinants of Resistance to Innovation: Peer Review, Disciplinary Paradigm, and Formal Organizations." Paper read at the annual meeting of the New York State Sociological Association, Potsdam, NY.

Scott, Marvin B., and Stanford M. Lyman. 1968. "Accounts." *American Sociological Review* 33 (1): 46–62.

Scott, Richard W., Martin Ruef, Peter J. Mendel, and Carol A. Caronna. 2000. *Institutional change and organizations: Transformation of Healthcare Organizations*. Chicago: University of Chicago Press.

Sexton, Donal J. 1986. "The Theory and Psychology of Military Deception." In *Deception: Perspectives on Human and Nonhuman Deceit*, edited by Robert W. Mitchelland Nicholas S. Thompson, 349–56. Albany: State University of New York Press.

Shapin, S. 1982. "The Sociology of Science." *History of Science* 20 (49): 157–211.

Sharp, D. W. 1991. "Fraud: The Journal's Role Concerning Fraudulent Research." *Investigative Radiology* 26 (6): 586–89.

Shenkman, Richard. 1993. *Legends, Lies & Cherished Myths of World History*. New York: HarperCollins.

Shermer, Michael. 2002. *Why People Believe Weird Things*. New York: A. W. H. Freeman/Owl Book, Henry Holt.

Sifakis, Carl. 1996. *The Big Book of Hoaxes*. New York: Paradox Press.

Silberner, J. 1982. "Cheating in the Labs." *Science Digest* 90 (8): 38, 40, 41.

Silk, Joseph. 2001. *The Big Bang*. New York: W. H. Freeman.

Simmons, Joseph P., Leif D. Nelson, and Uri Simonsohn. 2011. "False-Positive Psychology: Undisclosed Flexibility in Data Collection and Analysis Allows Presenting Anything as Significant." *Psychological Science* 22 (11): 1359–66 (see also http://papers.ssrn.com/sol3/papers.cfm?abstract_id=1850704).

Simpson, Sally S., and David Weisburd, eds. 2009. *The Criminology of White-Collar Crime*. New York: Springer.

Sinclair, David. 2004. *The Land that Never Was*. Cambridge, MA: Da Capo Press.

Singer, Barry, and Victor A. Benassi. 1981. "Occult Beliefs: Media Distortions, Social Uncertainty, and Deficiencies of Human Reasoning Seem to Be at the Basis of Occult Beliefs." *American Scientist* 69: 49–55.

Singh, Simon. 2004. *Big Bang. The Origin of the Universe*. New York: HarperCollins.

Sismondo, Sergio. 1996. *Science Without Myth. On Constructions, Reality, and Social Knowledge*. Albany: State University of New York Press.

Skogan, W. G. 1984. "Reporting Crimes to the Police: The Status of World Research." *Journal of Research in Crime and Delinquency* 21 (2): 113–37.

Smaglik, P. 2016. "Activism: Frustrated Postdocs Rise Up." *Nature* 530 (7591): 505–6.

Smith, William Jay. 1961. *The Spectra Hoax*. Middletown, CT: Wesleyan University Press.

Smyth, Denis. 2010. *Deathly Deception: The Real Story of Operation Mincemeat*. New York: Oxford University Press.

Snowdon, Christopher. 2010. *The Spirit Level Delusion: Fact-Checking the Left's New Theory of Everything*. London: Democracy Institute/Little Dice.

Sokal Hoax, The. 2000. Edited by the editors of *Lingua Franca*. Lincoln: University of Nebraska Press.

Sox, H. C., and D. Rennie. 2006. "Research Misconduct, Retraction, and Cleansing the Medical Literature: Lessons From the Poehlman Case." *Annals of Internal Medicine* 144 (8): 609–13.

Sprague, R. L. 1993. "Whistleblowing: A Very Unpleasant Avocation." *Ethics and Behavior* 3 (1):103–33.

St. James-Roberts, Ian. 1976. "Cheating in Science." *New Scientist* 72 (1028) (November 25): 466–69.

Steneck, N. H. 2006. "Fostering Integrity in Research: Definitions, Current Knowledge, and Future Directions." *Science and Engineering Ethics* 12 (1): 53–74.

Stewart, W. W., and N. Feder. 1987. "The Integrity of the Scientific Literature." *Nature* 325 (6101): 207–14.

Storer, Norman W. 1977. "The Sociological Context of the Velikovsky Controversy." In *Scientists Confront Velikovsky*, edited by D. Goldsmith, 29–39. New York and London: W.W. Norton.

Story, R. D. 1980. *The Encyclopedia of UFOs*. New York: Doubleday.

Swazey, J. P., M. S. Anderson, and K. S. Louis. 1993. "Ethical Problems in Academic Research." *American Scientist* 8: 542–53.

Swazey, J. P., and S. R. Scher, eds. 1981. *Whistleblowing in Biomedical Research: Policies and Procedures for Responding to Reports of Misconduct*. Washington, DC: U.S. Government Printing Office.

Sykes, G. M., and D. Matza. 1957. "Techniques of Neutralization: A Theory of Delinquency." *American Sociological Review* 22: 664–70.

Sykes, Steven, ed. 1991. *Deceivers Ever: Memoirs of a Camouflage Officer*. Speldhurst, UK: Spellmount.

Tammy L. M. 2008. "Framing and Organizational Misconduct: A Symbolic Interactionist Study." *Journal of Business Ethics* 78 (1–2): 3–16.

Tangney, June P. 1987. "Fraud Will Out—Or Will It? Scientists' Attitudes to Bogus Research." *New Scientist* 115 (1572): 62–63.

Thelen, M. H., and T. M. Di Lorenzo. 1992. "Academic Pressures." In *Research Fraud in the Behavioral and Biomedical Sciences*, edited by David J. Miller and Michel Hersen, 161–81 New York: John Wiley & Sons.

Thio, Alex. 1978. *Deviant Behavior*. Boston: Houghton Mifflin.

Tibballs, Geoff. 2006. *The World's Greatest Hoaxes*. New York: Barnes & Noble.

Tiryakian, Edward A. 1974. *On the Margin of the Visible: Sociology, the Esoteric and the Occult*. New York: John Wiley & Sons.

Toch, Hans. 1981. "Cast the First Stone? Ethics as a Weapon." *Criminology* 19 (2): 185–94.

Traill, David A. 1995. *Schliemann of Troy: Treasure and Deceit*. New York: St. Martin's Press.

Truzzi, Marcello. 1979. "On the Reception of Unconventional Scientific Claims." In *The Reception of Unconventional Science*, edited by S. H. Mauskopf. AAAS Selected Symposium 25. Boulder: Westview Press.

Tuomela, Raimo. 1987. "Science, Protoscience, and Pseudo Science." In *Rational Changes in Science*, edited by J. C. Pitt and M. Pera, 83–101. Boston Studies in the Philosophy of Science, vol. 98. Boston: D. Reidel.

Vaidyanathan, Brandon, Simranjit Khalsa, and Elaine Howard Ecklund. 2016. "Gossip as Social Control: Informal Sanctions on Ethical Violations in Scientific Workplaces." *Social Problems* 63 (4): 554–72.

Vasek, Marie M. 1986. "Lying as a Skill: The Development of Deception in Children." In *Deception: Perspectives on Human and Nonhuman Deceit*, edited by Robert W. Mitchell and Nicholas S. Thompson, 271–92. Albany: State University of New York Press.

Vaughan, Diane. 1998. "Rational Choice, Situated Action, and the Social Control of Organizations." *Law and Society Review* 32 (1): 23–61.

Vaughan, Diane. 1999. "The Dark Side of Organizations: Mistake, Misconduct, and Disaster." *Annual Review of Sociology* 25: 271–305.

Wade, Nicholas. 1975. "Discovery of Pulsars: A Graduate Student's Story." *Science* 189 (4200): 358–64.

Wade, Nicholas. 1976. "IQ and Heredity: Suspicion of Fraud Beclouds Classic Experiment." *Science* 194: 916–19.

Wade, Nicholas. 1981. "A Diversion of the Quest for Truth." *Science* 211 (4486) (March 6): 1022–25.

Wade, Nicholas. 1983. "Fraud and the State of Science." *SUNY Research* 2 (4): 3–6.

Wagner, Caroline S. 2008. *The New Invisible College: Science for Development*. Washington, DC: Brookings Institution Press.

Walker, Mark. 1995. *Nazi Science*. Cambridge, MA: Perseus.

Wallis, Claudia. 1983. "Fraud in a Harvard Lab." *Time*, February 28 (International Edition).

Wallis, Roy. 1979. *On the Margins of Science: The Social Construction of Rejected Knowledge*. Staffordshire, UK: University of Keele Press.

Weber-Wulff, Debora. 2014. *False Feathers: A Perspective on Academic Plagiarism*. Berlin: Springer.

Weick, K. E. 2013. Book review of Donald Palmer, *Normal Organizational Wrongdoing: A Critical Analysis of Theories of Misconduct in and by Organizations. Administrative Science Quarterly* 58 (2): 292–95.

Weiner, S. J. 1980. *The Piltdown Forgery*. New York: Dover.

Weinstein, Deena. 1978. *Fraud in Science*. A paper given at the 73rd Annual Meeting of the American Sociological Association, San Francisco.

Weinstein, Deena. 1979. "Fraud in Science." *Social Science Quarterly* 59 (4): 639–52.

Weisburd, David, Stanton Wheeler, Elin Waring, and Nancy Bode. 1991. *Crimes of the Middle Classes: White-Collar Offenders in the Federal Courts*. New Haven: Yale University Press.

Weisburd, David, Elin Waring, and Ellen F. Chayet. 2001. *White-Collar Crime and Criminal Careers*. New York: Cambridge University Press.

Weisman, Richard. 2014. *Showing Remorse: Law and the Social Control of Emotion*. Surrey: Ashgate.

Wells, Don, Alison J. Donnell, Adrian Thomas, Melanie Sandifer Mills, and Mark Miller. 2006. "Creative Deviance: A Study of the Relationship between Creative Behavior and the Social Construct of Deviance." *College Student Journal* 40 (1) (March): 74–77.

Westfall, R. S. 1973. "Newton and the Fudge Factor." *Science* 179 (February 23): 751–58.

Wheatley, Dennis. 1980. *The Deception Planners*. London: Hutchinson.

Wheeler, S., D. Weisburd, and N. Bode. 1982. "Sentencing the White Collar Offender: Rhetoric and Reality." *American Journal of Sociology* 47 (5): 641–59.

Wicherts, J. M., D. Borsboom, J. Kats, and D. Molenaar. 2006. "The Poor Availability of Psychological Research Data for Reanalysis." *American Psychologist* 61 (7): 726–28.

Wiener, Jon. 2005. *Historians in Trouble: Plagiarism, Fraud, and Politics in the Ivory Tower*. New York: New Press.

Wilkinson, Richard, and Kate Pickett. (2009) 2011. *The Spirit Level: Why Greater Equality Makes Societies Stronger*. New York: Bloomsbury Press.

Wilson, Edward O. 1998. *Consilience: The Unity of Knowledge*. New York: Knopf.

Wolins, L. 1962. "Responsibility for Raw Data." *American Psychologist* 17:657–58.

Woolf, Patricia. 1981. "Fraud in Science: How Much, How Serious?" *Hastings Center Report* 11 (5): 9–14.

Woolf, Patricia. 1988. "Deception in Scientific Research." *Jurimetrics Journal* 29 (1): 67–95.

Wright, S. 1966. "Mendel's Ratios." In *The Origins of Genetics: A Mendel Sourcebook*, edited by Curt Stern and Eva R. Sherwood, 173–75. San Francisco: W. H. Freeman.

Wynne, Brian. 1979. "Between Orthodoxy and Oblivion: The Normalization of Deviance in Science." In *On the Margins of Science*, edited by R. Wallis, 67–83. Staffordshire, UK: University of Keele.

Young, Martin, and Robbie Stamp. 1989. *Trojan Horses: Deception Operations in the Second World War*. London: Mandarin.

Zagorin, Perez. 1996. "The Historical Significance of Lying and Dissimulation." *Social Research* 63 (3): 863–912.

Zebrowitz, Leslie. 1997. *Reading Faces: Windows to the Soul*. Boulder: Westview Press.

Zimbardo, Philip G. 1973. "A Study of Prisoners and Guards in a Simulated Prison." *Naval Research Reviews* 9 (1): 1–17.

Zinder, Norton D. 1983. "Fraud in Science." *Science '83* 4 (1): 94–95.

Zuckerman, Harriet. 1968. "Patterns of Name-Ordering Among Authors of Scientific Papers." *American Journal of Sociology* 74:76–291.

Zuckerman, Harriet. 1977. "Deviant Behaviour and Social Control in Science." In *Deviance and Social Change*, edited by E. Sagarin, 87–138. Beverly Hills, CA: Sage Publications.

Zurer, Pamela S. 1987. "Misconduct in Research: May It Be More Widespread than Chemists Like to Think?" *Chemical and Engineering News* 65 (15): 10–17.

Index

Printed and bound by CPI Group (UK) Ltd, Croydon, CR0 4YY

16/09/2026

14963487-0004